Research Experiences in Plant Physiology

Second Edition

Thomas C. Moore

Research Experiences in Plant Physiology

A Laboratory Manual

Second Edition

With 23 Figures

Springer-Verlag   New York Heidelberg Berlin

Thomas C. Moore
Professor of Botany
Department of Botany and Plant Pathology
Oregon State University
Corvallis, Oregon 97331
U.S.A.

Printed in the United States of America.

9 8 7 6 5 4 3 2 1

ISBN 0-387-90606-1 Springer-Verlag New York Heidelberg Berlin
ISBN 3-540-90606-1 Springer-Verlag Berlin Heidelberg New York

Preface to the Second Edition

Three major changes have been made in *Research Experiences in Plant Physiology* in producing this second edition. The format has been changed so as to minimize the number of pages and hence the cost to students, hopefully without sacrifice of readability or general utility of the manual. Three new exercises have been added on Phytochrome Effects in Nyctinastic Leaf Movements (Exercise 26), Measurement of Leaf Water Potentials with a Pressure Chamber (Exercise 27), and The Hill Reaction of Photosynthesis (Exercise 28) in an effort to provide more balanced coverage of the major topics in Plant Physiology. Lastly, modest revisions have been made in the text and lists of references throughout the manual and in the index. It is hoped that these collective changes will lead to continued wide acceptance of *Research Experiences in Plant Physiology* as the laboratory manual for upper-division undergraduate and graduate courses in Plant Physiology.

The preparation of this new edition naturally has involved the invaluable assistance of several persons. I owe special thanks to Mrs. Ellen Witt for her patience and proficiency in retyping the entire manual. To Mrs. Witt, Mrs. Leona Nicholson, and my wife, Arvida, I am grateful for assistance in proofreading. I thank Brian D. Cleary for assisting with the writing of Exercise 27 on leaf water potentials, and Donald J. Armstrong for his valuable criticism and suggestions regarding many of the exercises. Finally, I thank Mr. Stephen J. Danko for assisting with the testing of the new exercises.

Thomas C. Moore
Corvallis, Oregon
February 1981

Preface to Instructors

Research Experiences in Plant Physiology is intended as a laboratory manual for an upper-division undergraduate or graduate course in Plant Physiology. Depending upon the length and number of laboratory sessions per week, the manual is suitable for courses of one quarter to one year in duration. It is expected that students using the manual will have had among their previous college-level courses the equivalent of one year of General Biology or General Botany, one year of General Chemistry, and at least an introductory course in Organic Chemistry or Biochemistry, or that they will be taking the latter Chemistry concurrently.

The manual has developed over a period of 12 years during which I have taught over 20 classes, and some 40 to 50 separate laboratory sections, in three different undergraduate courses in Plant Physiology. The enrollment per laboratory section has varied from 12 to 24 students, with each student attending either one or two three-hour sessions per week, depending upon the particular course. Always it has been my practice, by personal preference, to directly supervise each laboratory session with the help of a graduate teaching assistant.

Included are exercises pertaining to most major areas of Plant Physiology, although definite emphasis is placed on growth and development of angiosperms. In the case of a few topics—photosynthesis, dormancy, growth regulators—two or more related exercises are presented. I do not necessarily advocate that two closely related exercises be conducted by the same students in a given course. Rather, my reason for including related exercises on some topics is that I believe it is often advantageous to have different laboratory sections in one course do different but somewhat analogous exercises, and to vary the specific exercises assigned each time the course is offered. A concerted effort has been made to avoid duplication of exercises that a majority of students are likely to do in other courses and to emphasize topics unique to the physiology of angiosperms. Thus, purposely excluded are several exercises which normally are parts of an undergraduate course in Cell Physiology. Some other exercises which reasonably might have been added, but which I believe to be more appropriately left to a Biochemistry course, also are purposely excluded. These considerations, together with the unavoidable influence of my own specialization, account for the emphasis on growth and development. Other users must, of course, judge for themselves whether the exercises selected for inclusion are adequately representative and whether important topics have been injudiciously omitted.

Each exercise in *Research Experiences in Plant Physiology* is complete in itself, with no reference to appendices and, except in a very few cases, other exercises required. Two advantages in particular accrue from the adoption of this format. The chief advantage is that instructors who do not wish to adopt the whole manual can utilize separately or in specific combination only certain exercises. A second advantage is that I can expeditiously revise the manual frequently by adding and deleting exercises, as well as modifying those now included.

Each exercise is written in five sections. The first is an ''Introduction'' the intended purpose of which is to acquaint the student with the area of investigation and to provide him with the rationale, context and perspective for the prescribed experiment or other procedure. The difficulty of closely coordinating lectures with laboratory exercises which sometimes are several weeks in duration is well known to instructors in Plant Physiology courses. Hopefully, the introductory narratives will aid in circumventing that problem.

A "Materials and Methods" section follows next in each exercise. Here, of course, the actual prescribed procedure is described in detail. Preparations of special reagents and plant materials are described, sources of some special items are indicated, descriptions of special apparatus are included, and, in some exercises, alternative procedures and satisfactory substitutions are specified. To avoid making the narrative excessively difficult to read, extensive use is made of footnotes in this section. Some difficulties are certain to arise concerning scheduling of particular exercises, particularly in courses in which there is only one laboratory session per week. In such courses, it will be inescapably necessary for students to report to the laboratory a few times outside the class schedule. Where students attend two laboratories per week such problems are so infrequent as to be negligible. Instructors are urged to check this section carefully to select the actual step in the procedure where they wish the students to begin. For some exercises it is advisable—in a few it is imperative—to have plants of a particular age, size or condition ready for presentation to the class at the time experimental manipulations are to be started. Some instructors will wish to have most reagents prepared in advance; others will prefer to have the students prepare most or all reagents.

Next is a rather extensive list of references in each exercise. Generally, students are not expected to read or even to consult all the references listed in each exercise. The list is made fairly extensive, however, for the benefit of students who happen to be particularly interested in that exercise or area of investigation. For those students the list of references usually will provide a good entry into the generally extensive literature on each topic.

The fourth section of each exercise is a list of the special reagents, supplies and equipment required. The list does not necessarily include common items which are assumed to be generally present in the laboratory and for which no instructions for preparation need be specified. In the subheading denoting each list, I have indicated my recommendation as to how the procedure is to be handled, whether by individual students, by teams of students (if by teams, how many students in each), as a class project, or as a demonstration.

Implicit in the recommendation regarding procedural handling of each exercise is one of my strongest personal pedagogical prejudices. I am convinced that laboratory exercises should provide students direct experience in actual research-type investigations. Each exercise should, I believe, be sufficiently comprehensive in design and include adequate replication to be realistically instructive about the principle or process under study or the technique being practiced. At the same time, I feel compelled to have students perform a diversity of exercises adequate to cover most major topics in Plant Physiology. The best approach toward maximum realization of these dual objectives, which I have found, is to have students work in teams. In my own laboratories most of the exercises are performed by teams of three or four students, some by teams of as many as eight students, and a few are conducted by all the students in a section cooperating as one team. Occasionally, the separate teams in a section do different, though closely related experiments, and the results of the collective experiments are observed and discussed by all the students. I readily concede that this practice is not without disadvantages, the most serious being that some students tend to exclude themselves from active participation and hence learn less than if they worked independently. On balance, however, I firmly believe that the advantages exceed the disadvantages.

The final section of each exercise is a set of removable report forms. These include skeleton tables and graphs and blank spaces for recording data and observations. A separate form contains questions with spaces provided for answers. Some instructors undoubtedly will wish to make only limited use of these report forms, preferring instead that students write independent reports—in the form of a technical paper—on at least a few exercises. This always has been my routine practice, because of the obvious benefit students derive from the writing experience, which, of course, is but the final step in any significant research project. For the benefit of those who elect this option, I have included a set of instructions for the preparation of such independent reports in the front of the manual. Even in those cases where independent reports are written, the report forms will serve as guides to students as to the types of information which should be covered in the reports.

The origins of the exercises in *Research Experiences in Plant Physiology* naturally are varied. Many were adapted from research papers which are, in each case, acknowledged.

Whenever possible, one or more authors of each publication used in this way was consulted and criticisms of the exercise invited.

Several of my colleagues at Oregon State University have made very substantial contributions to the manual. Ralph S. Quatrano contributed the exercise on "Polyacrylamide Gel Electrophoresis of Plant Proteins." "Symbiotic Nitrogen Fixation in Legume Nodules" was contributed by Miss Kathleen Fishbeck and Harold J. Evans. Assistance in the development of the exercise on "Potassium Activation of Pyruvic Kinase" was generously given by Harold J. Evans, from whose publications the exercise was directly adapted. Norman I. Bishop contributed the basic procedures employed in the two exercises on photosynthesis and advised me on procedures for the thin-layer chromatography of chloroplast pigments. I developed the remainder of the exercises, most of them from research publications by my associates and me, some from our unpublished investigations.

Acknowledgments

To all persons mentioned above I extend my sincere gratitude for their many valuable contributions. For their reviews and constructive advice regarding particular exercises based on their research papers, appreciation also is expressed to John T. Barber, Bryan G. Coombe, John D. Goeschl, Masayuki Katsumi, A. Carl Leopold, J. van Overbeek, Leslie G. Paleg, Tsvi Sachs, Ernest Sondheimer, Kenneth V. Thimann, and Israel Zelitch. Special thanks are due Miss Kathleen Fishbeck, not only for her contribution of one exercise, but also for her invaluable assistance with the preparation and editing of the manuscript. Ronald C. Coolbaugh, Paul R. Ecklund, Harold J. Evans, Ralph S. Quatrano, Patricia T. Stevens and Donald B. Zobel generously gave of their time to review parts or all of the final draft of the manuscript. Mrs. Susan A. Barlow assisted greatly with numerous tasks in the final preparation of the manuscript. To Mrs. Linda K. Fletcher and Mrs. Mary Ellen Witt I express sincere thanks for their patient and skillful typing of the manuscript. The assistance of Miss Dianne Baumann with typing of draft copy also is gratefully acknowledged. Credit for drawing all the original illustrations except Figure 16-1 goes to Mr. James Wong. Finally, I express sincere appreciation to my former professors and the many students and teaching assistants who inspired me to undertake the writing of this manual. In this regard, I extend special personal acknowledgment to Archibald W. Roach, my first botany professor, whose zeal for botany I found contagious as a college freshman nearly twenty years ago.

While I gratefully acknowledge the many persons who have contributed to the publication of *Research Experiences in Plant Physiology*, I alone am responsible for any errors or inadvertent improprieties that may have escaped recognition.

Thomas C. Moore
Corvallis, Oregon

Preface to Students

Welcome! Presumably you have just enrolled in your first course in Plant Physiology or will do so in the near future. Whether the course is a required one or an elective for you, I believe I can assure you that you are embarking upon a challenging and gratifying educational venture.

The exercises contained in this manual are intended to be, as the title indicates, *research experiences* for you. I hope that in this laboratory you will feel like a professional plant physiologist does in the daily pursuit of his research interests. Hopefully, you will be challenged in your efforts to execute sometimes quite complex procedures successfully and in analyzing and interpreting the results of your efforts. Hopefully, you will be excited about your work and will be eager to learn the outcome of your experiments. Hopefully, you will take pride in having successfully completed an instructive experiment. Hopefully, you will learn to work efficiently and effectively with your peers as a member of a research team. Hopefully, you will rebound from the failures and disappointments that you almost certainly will infrequently experience. This, after all, is also part of the real-life experience of every professional researcher. And, hopefully above all, you will *learn by doing* direct experimental investigations much about the fascinating functions of plants.

I suggest that you adopt as routine practice familiarizing yourself with each laboratory assignment in advance of the period in which the tasks are to be performed. Discover what the exercise is all about, how it is to be conducted, and what kinds of data are required to be taken and observations to be made. Upon terminating an exercise, summarize the results and work up your report promptly.

The instructional staff will guide you regarding matters of safety in the laboratory. The "Materials and Methods" sections of the exercises also contain frequent notes of caution. But you personally must accept a large measure of responsibility for your own and your fellow students' personal safety. In general, remember and practice the rules of safety which you learned in your previous chemistry courses. Beyond that, always ask an instructor in advance whenever you are in doubt about how to handle a particular reagent or use a particular item of equipment. Be alert to potential hazards which may be created by others working around you.

Best wishes for a successful laboratory.

Thomas C. Moore

Contents

Instructions for Writing Laboratory Reports in the Form of Technical Papers[1]

An essential part of a significant scientific investigation is communication of the findings to others interested in the particular field. Hence, preparation of at least some of your laboratory reports in the form of a scientific paper is regarded as an important part of your experience in this course.

The particular format and style adopted for a given paper depends upon a number of factors, particularly the precise nature of the report and the journal or other publication in which the paper is to be published. Each periodical journal or other type of publication has adopted a particular format and style and demands of authors rather strict conformity to its practices. An author must, therefore, discipline himself to write each paper in a form acceptable to the journal in which he expects to publish his work.

In this course, you are expected to follow the format and style of the journal _Plant Physiology_, except for minor details in the listing of references, in the preparation of your laboratory reports. The practices adopted by this journal are not necessarily the best ones, and by no means are they the only acceptable practices. However, in the interest of uniformity and discipline, you are instructed to follow generally the practices of _Plant Physiology_. You are advised to become familiar with details of organization, section headings, methods of presentation of data, and ways of citing references by examining papers in recent issues of the journal. There follows below a brief guide, set up by section, which should be helpful to you. Reports should be typed, if possible, and typing should be double-spaced throughout the report.

Title (Capitalize only the first letter of each major word.)

AUTHOR

ABSTRACT

This section should be a very brief condensation of the whole paper; that is, a brief statement of what was done, by what general procedures it was done, the main findings, and the conclusions drawn from the results. Limit your conclusions to those statements which can be inferred directly from the results of your investigation. Write the "abstract" in the past tense.

---

[1]Alternative to the use of printed Report forms supplied with each exercise.

Introduction (Section heading is not used.)

This brief section should contain: (1) a description of the nature of the problem and the state of the problem at the beginning of the investigation; (2) a statement of the purpose, scope and general method of the investigation; and (3) a brief statement of the most significant findings of the investigation. For the purpose of laboratory reports, only items 2 and 3 need to be emphasized.

MATERIALS AND METHODS

Describe in detail the materials and equipment employed and the methods followed in the investigation. This section should be written so completely and explicitly that a competent investigator could repeat your experiments. In the preparation of an actual scientific paper, this section would be written in the past tense. For your laboratory reports you may, if you wish, simply copy or paraphrase the "Materials and Methods" sections of the laboratory exercises.

RESULTS

Observations and experimental data are recorded in this section. An important recommendation is to graph the data whenever possible, because relationships are much more readily visualized in a graphic presentation than in a table. And remember that some data which cannot be graphed on coordinates can, nevertheless, be expressed in pictorial graphic form in a histogram. Write in the past tense.

DISCUSSION

Discuss the main findings of the investigation. Present the evidence for each conclusion. Try to account for unexpected results and exceptions. Compare your results and interpretations with those of other professional investigators and other members of the class. Obviously, this section should include pertinent information from reference books, periodical literature, and other sources, so as to indicate the significance of the process or phenomenon studied and to relate your results to the findings of others. The questions contained in the Report forms generally serve as guides to the types of information which should be incorporated in this section.

LITERATURE CITED

This is the last section of a scientific paper. References are listed alphabetically by author and are numbered consecutively, as indicated by the sample list below. Papers are referred to in the paper by number or by author(s) and number, as the examples below indicate.

Format for Listing References

1. Addicott, F. T. and J. L. Lyon. 1969. Physiology of abscisic acid and related substances. Ann. Rev. Plant Physiol. 20: 139-164.

2. Briggs, W. R., W. D. Zollinger, and B. B. Platz. 1968. Some properties of phytochrome from dark-grown oat seedlings (Avena sativa L.). Plant Physiol. 43: 1239-1243.

3. Fox, J. E. 1969. The cytokinins. Pp. 85-123 in: M. B. Wilkins, Ed. The Physiology of Plant Growth and Development. McGraw-Hill Book Company, New York.

4. Salisbury, F. B. and C. W. Ross. 1978. Plant Physiology. 2nd Ed. Wadsworth Publishing Company, Belmont, California.

Methods for Citing References in Text

Fox (3) discussed structure-activity relationships for the cytokinin-type growth substances.

It is now firmly established that abscisic acid is a natural plant hormone (1).

Briggs et al. (2) described some properties of phytochrome from etiolated Avena seedlings.

The mechanisms of action of the plant hormones are incompletely understood at present (1, 3, 4).

Chemical Composition of Cell Membranes and Factors Affecting Permeability

Introduction

Every cell possesses an elaborate and complex system of membranes, the general function of which is to regulate exchange of materials between the cell and its surroundings and among the organelles of the cell and to provide for compartmentation of cellular metabolites. Thus a generalized type of higher plant cell has a <u>plasmalemma</u> (or plasma membrane), which constitutes the outer limiting boundary of the protoplast next to the cell wall; a <u>tonoplast</u> (or vacuolar membrane), which separates the contents of each vacuole (normally one large vacuole in a mature higher plant cell) from the cytoplasm; and membranes bounding the various organelles, including the <u>chloroplasts</u> (if present), <u>mitochondria</u>, <u>endoplasmic reticulum</u>, <u>dictyosomes</u>, <u>peroxisomes</u> and <u>glyoxysomes</u> (if present). Some organelles, specifically chloroplasts and mitochondria, are actually bounded by double membranes, with the inner membrane showing elaborate organization. The nuclear envelope, consisting of double membranes, possesses pores of quite sizable diameter (500–1000Å).

Investigations of the chemical composition, ultrastructure and functions of membranes have provided conclusive evidence that all the membrane fractions of cells possess certain features in common. For one thing, membranes generally have been found to consist of approximately <u>60% protein and 40% lipids in gross chemical composition</u>. The nature of the structural protein is largely unknown, but specific enzymes have been shown to be associated with, and apparently to be integral parts of, particular membranes. The lipids appear to be mainly phospholipids, glycolipids, sulfolipids and sterols and other complex lipids. The specific amounts and kinds of lipids vary widely among different types of membranes, with each type of membrane displaying a unique composition.

Information regarding the molecular architecture of membranes and the types of forces causing the cohesion of the protein and lipoidal constituents is very incomplete. Theoretical models of membrane molecular architecture are of 3 general types: bi-molecular layer of lipid interposed between mono-molecular layers of protein; mosaic of lipoidal areas within a continuous protein mesh; or repeating subunits of protein-lipid composition. The cohesive forces binding the constituents appear to be predominantly hydrophobic bonds.

A second property, besides similarity of gross chemical composition, which membranes generally have in common is the very important functional property of <u>differential permeability</u>. Membranes permit water to pass through them with relative ease with net movements of water being passive and in strict accordance with gradients in water potential. Membranes also permit a variety of solutes to pass through them but at markedly different rates and by different mechanisms, while, of course, totally excluding passage of some materials.

The purpose of this experiment is to investigate the effects of various chemical and physical treatments on membrane permeability, and to interpret those effects on the basis of the known gross chemical composition of membranes generally. The plant material selected for study is the storage root of the common garden beet (Beta vulgaris), the cells of which contain conspicuous amounts of red-purple pigment, betacyanin, dissolved in the cell sap of the vacuoles. The tonoplast and plasmalemma are normally essentially impermeable to betacyanin. However, if a treatment adequately increases permeability, the pigment can be observed to leak out of the beet root cells. Hence the effects of various treatments on increasing membrane permeability can be quantified by measuring the amount of pigment leaking out of a sample of tissue.

Materials and Methods

Each group of students select one large beet root and wash it thoroughly with a brush and tap water. Using a cork borer of approximately 1 cm diameter, cut 12 cylinders from the same root if possible. Cut all the cylinders to a uniform length of approximately 3 cm, leaving a clean transverse cut at each end. Wash all the cylinders in running tap water for about 10 to 15 minutes, in order to remove pigment on the surfaces. Then perform each of the following treatments.

Heat treatment. Prepare a water bath by placing a 600-ml or 1-liter beaker about two-thirds filled with distilled water on a hot plate or over a burner. Using a clamp and ring stand, rig up a thermometer so as to have the bulb suspended in the water. Heat the water bath to 70° C and submerse one cylinder in the water bath for precisely 1 minute. The cylinder may be held with forceps or impaled on a dissection needle. Then transfer the cylinder to a test tube containing 15 ml of distilled water at room temperature. Now remove the source of heat and allow the water bath to cool naturally, treating one cylinder each also at 65°, 60°, 55°, 50° and 45°. Place one cylinder directly in a test tube containing 15 ml of distilled water (measure and record temperature) to serve as a control. After one hour of incubation of each cylinder, shake the contents of the test tube, pour a sample of the bathing solution into a colorimeter tube and measure the percent absorption of light at 525 nm on a colorimeter. Note: It is advisable to save all the samples of incubation solution until the solution containing the highest concentration of pigment has been analyzed. If this solution is too optically dense to get an accurate measurement, dilute all samples identically (e.g. 1:1), with distilled water and repeat the measurements.

Freezing treatment. Cut a cylinder of identical size to those used in the preceding part from a beet root which has been previously quick-frozen and allowed to thaw. Rinse the section very quickly in tap water and transfer it to a test tube containing 15 ml of distilled water. Place a second, control cylinder directly in another test tube with 15 ml of distilled water. After incubation for one hour, measure the relative amount of pigment in the solution surrounding each section. Note: If solutions were diluted in the first part of the experiment, these solutions should be diluted also in the same manner.

Treatment with organic solvents. Place 1 cylinder in 15 ml of each of the following solvents: methanol, acetone, benzene. Place another cylinder in 15 ml of distilled water. Incubate for one hour and measure the relative amount of pigment that leaks out of each sample. If samples are diluted, dilute by same factor as in first two parts of the experiment, using the appropriate solvent in each case. Notes: It will be necessary to compare each of these incubation solutions with the appropriate solvent as a blank in the colorimetry. The benzene sample will be difficult to measure, because there will be formed an unstable emulsion of water droplets containing pigment in benzene upon shaking the contents of the test tube; shake the solution and take the measurement as quickly as possible.

As an optional addition to this experiment, designed to assist in interpreting the effects of the organic solvents on membrane permeability, approximately equal quantities of each solvent can be mixed separately with fresh egg white (a protein) and olive oil (a simple lipid, tri-glyceride) and shaken up and the reaction, if any, detected visually.

References

1. Branton, D. 1969. Membrane structure. Ann. Rev. Plant Physiol. 20: 209-238.

2. Collander, R. 1957. Permeability of plant cells. Ann. Rev. Plant Physiol. 8: 335-348.

3. Frey-Wyssling, A. and K. Mühlethaler. 1965. Ultrastructural Plant Cytology, with an Introduction to Molecular Biology. Elsevier Publishing Company, Amsterdam.

4. Giese, A. C. 1968. Cell Physiology. 3rd Ed. W. B. Saunders Company, Philadelphia.

5. Green, D. E. and J. F. Perdue. 1966. Membranes as expressions of repeating units. Proc. Nat. Acad. Sci. 55: 1295-1302.

6. Jensen, W. A. 1970. The Plant Cell. 2nd Ed. Wadsworth Publishing Company, Belmont, California.

7. Jensen, W. A. and R. B. Park. 1967. Cell Ultrastructure. Wadsworth Publishing Company, Belmont, California.

8. Korn, E. D. 1966. Structure of biological membranes. Science 153: 1491-1498.

9. Ledbetter, M. C. and K. R. Porter. 1970. Introduction to the Fine Structure of Plant Cells. Springer-Verlag, New York.

10. Ray, P. M. 1972. The Living Plant. 2nd Ed. Holt, Rinehart and Winston, New York.

11. Rothfield, L. and A. Finkelstein, 1968. Membrane biochemistry.
    Ann. Rev. Biochem. 37: 463-496.

12. Ruesink, A. W. 1971. The plasma membrane of Avena coleoptile
    protoplasts. Plant Physiol. 47: 192-195.

13. Salisbury, F. B. and C. W. Ross. 1978. Plant Physiology. 2nd Ed.
    Wadsworth Publishing Company, Belmont, California.

14. Siegel, S. M. and O. Daly. 1966. Regulation of betacyanin
    efflux from beet root by poly-L-lysine, Ca-ion and other
    substances. Plant Physiol. 41: 1429-1434.

15. Troshin, A. S. 1966. Problems of Cell Permeability.
    Pergamon Press, Oxford.

Special Materials and Equipment Required
Per Team of 3-4 Students

(1) Large garden beet (Beta vulgaris) root (local market)
(1) Beet previously frozen and thawed at beginning of laboratory
(1) Cork borer approximately 1 cm in diameter
(Approximately 50 ml each) Methanol, acetone and benzene
(1) Colorimeter or spectrophotometer and cuvettes

Recommendations for Scheduling

It is suggested that students work in teams of 3 to 4 members, each team performing the entire experiment in a single laboratory period.

Duration: 1 laboratory period; 2 to 3 hours exclusive attention to this experiment required.

EXERCISE 1

CHEMICAL COMPOSITION OF CELL MEMBRANES AND FACTORS AFFECTING PERMEABILITY

REPORT

Name _____ Section _____ Date _____

Results of heat treatment:

Temperature (°C)

*Temperature of distilled water directly out of tap.

Results of freezing treatment:

Tissue sample	% Absorption at 525 nm
Untreated	_____
Frozen and thawed	_____

Results of treatment with organic solvents:

Solvent	% Absorption at 525 nm
Distilled water	_____
Methanol	_____
Acetone	_____
Benzene	_____

EXERCISE 1

CHEMICAL COMPOSITION OF CELL MEMBRANES AND
FACTORS AFFECTING PERMEABILITY

Report

Name _____ Section _____ Date _____

Questions

1. How do the results of the heat treatment, when plotted graphically, compare generally with thermal denaturation curves for proteins?

2. Explain the results with the quick-frozen cylinder of beet root. Briefly discuss the conditions under which freezing injury of plant tissues is due to intracellular ice crystal formation and laceration of cells and to desiccation.

3. Explain the effect of each of the organic solvents tested, in view of what is known about the gross chemical composition of membranes.

4. Outline briefly the features of solutes (e.g. lipoid solubility, degree of ionization, molecular and ionic size) which are of primary importance in determining their relative rates of penetration through cell membranes.

EXERCISE 2

MEASUREMENT OF THE WATER POTENTIAL OF PLANT TISSUES

General Introduction

Maintenance of the physiologically active state in individual cells
and whole multicellular plants is dependent upon relative constancy of a
number of conditions, one of which is favorable water balance. When, as
a part of the normal course of development or because of an inadequate
supply of water, plants have a reduced water content, the rate of their
development and, in general, the rates of all vital functions proceed at
reduced rates. And, of course, extreme or prolonged desiccation is
lethal to an actively growing plant.

Although, in the vigorously growing plant, development is impeded by
reduced water content of the tissues, the desiccated state is, on the
other hand, of great positive importance for the survival of the plant.
Thus, although dry seeds will not develop, neither are they killed by
high or low temperatures that would be lethal to the vegetative plant.
In fact, adaptation of plants to both drought and low temperature frequently involves a reduction in water content.

The purpose of this exercise is to learn about the physical principles
governing net water fluxes in osmotic systems and to become familiar with
one popular method for measuring the water potential of plant tissues.
The information with which this elementary exercise is concerned is basic
to understanding all net fluxes of water not only across membranes of
single cells, but from cell to cell and via vascular tissues within the
plant, from soil to plant, and plant to atmosphere. Thus the ultimate
purpose of the exercise goes well beyond the relatively simple system
selected for study.

Plant Cells as Osmometers

In plant cells (indeed all cells) there are many restrictions,
imposed by differential permeability of membranes, on the free diffusion
of solutes. There is a plasmalemma or plasma membrane forming the outer
boundary of the protoplast and an elaborate intracellular membrane system
forming barriers about each of the organelles (e.g. chloroplasts and
mitochondria) and forming a barrier (the tonoplast or vacuolar membrane)
between each vacuole and the cytoplasm. All these membranes possess the
property of differential permeability, which means that water molecules
move through them passively with relative ease but that solute molecules
and ions pass through them at different rates and often by other than
passive processes.

Thus, the essential features of an osmotic system are present in the
plant cell. These essential features are two. One, solutions or pure
water must be isolated by a membrane which restricts the passage of
solute particles more than it restricts passage of solvent particles.
Two, an osmometer includes some means of allowing pressure to build up

internally; in the plant cell, pressure can readily be built up because of the relative rigidity of the cell wall.

Since the vacuole constitutes the bulk of the volume of most mature plant cells, the vacuole plays a dominant role in the cell's osmotic water balance. The consequences of osmosis in a plant include not only changes of pressure within cells but also bulk movement of solutions, via the vascular tissues, within the plant.

Concept of Water Potential

In order to fully understand water relations, it is necessary to have some familiarity with some principles of thermodynamics. _Thermodynamics_ is the science of energy changes occurring in physical and chemical processes, including, of course, those occurring in cells.

At the outset, one should understand what is meant by _free energy_ or _Gibbs free energy_ (G). The free energy (G) is a thermodynamic property of a system or a component of a system and is defined as the energy isothermally (at a constant temperature) available for conversion to work. It may be defined also by the following equation:

$$G = E + PV - TS$$

where:

E = internal energy (the sum of the translational, rotational and vibrational energies of the substance).

PV = the pressure-volume product. If P is expressed in atmospheres and V in liters, these may be converted to calories, since one liter-atmosphere equals 24.2 calories.

T = absolute temperature (0° C = 273° T).

S = entropy (measure of the randomness; has units of energy per degree, calories/degree).

The free energy of a substance in any system is dependent upon the _amount_ of substance present; that is, upon the number of particles having a particular energy and entropy under the defined conditions of temperature and pressure. Free energy, therefore, is usually stated in terms of _energy per mole_ or per gram of the substance in question, e.g. calories/mole.

The _free energy per mole_ of any particular chemical species (e.g. water) in a multicomponent system (e.g. a solution) is defined as the chemical _potential_ of that species. The larger the chemical potential of a substance, the greater is its tendency to undergo chemical reactions and other processes such as diffusion.

12

The chemical potential of water is referred to as water potential (Ψ, Psi) and is a property of great importance to an understanding of water movement (net fluxes) in the plant-soil-air system. Water potential (Ψ), while sometimes expressed in energy terms (e.g. calories/mole), is usually expressed in terms of pressure (e.g. bars). Regardless of how it is expressed, if Ψ differs in various parts of a system, water will tend to move to the point where Ψ is lowest. Thus diffusion, including osmosis, occurs in response to a gradient in the free energy of diffusing particles.

Absolute values of chemical potential or of water potential (Ψ) are not easily measured, but differences in Ψ can be measured. The standard reference is conventionally taken to be the Ψ of pure water. Hence, water potential (Ψ) is the difference in free energy or chemical potential per unit molal volume between pure water and water in cells at the same temperature. The Ψ of pure water at atmospheric pressure is equal to zero; hence the Ψ of water in cells and solutions is typically less than zero, or negative.

The Ψ is affected by all factors which change the free energy or chemical activity of water molecules. Thus, Ψ is increased (made less negative) by:

1. Development of (turgor) pressure.
2. Increase in temperature.

And Ψ is decreased (made more negative) by:

1. Addition of solutes.
2. Matric forces which adsorb or bind water.
3. Negative pressures (tensions).
4. Reduction in temperature.

Ψ is an expression of the free energy status of water. It is a measure of the driving force which causes water to move into any system, such as plant tissue, soil or atmosphere, or from one part of the system to another. Ψ is probably the most meaningful property that can be measured in the soil-plant-air system. Ψ is the determinant of diffusional water movement, and bulk movement also occurs in response to gradients set up by diffusional movement under the control of a water-potential gradient.

Components of Cell Water Potential (Ψ)

The fundamental expression illustrating the components of Ψ is:

$$\Psi_{cell} = \Psi_\pi + \Psi_p + \Psi_m{}^1$$

where:

Ψ_{cell} = water potential of a cell.
Ψ_π (or Ψ_s) = osmotic potential.
Ψ_p = pressure potential (turgor pressure).
Ψ_m = matric potential.

Osmotic potential (Ψ_π or Ψ_s) is the contribution made by dissolved solutes to Ψ. It is always negative in sign.

Pressure potential (Ψ_p) is the contribution made by pressure to Ψ. It may be any value, either positive, zero, or negative. Addition of pressure (development of turgor pressure) results in a positive Ψ_p. Development of a tension (negative turgor pressure) in a living cell is a rare occurrence.

Matric potential (Ψ_m) is the contribution made by water-binding colloids and surfaces (e.g. cell walls) in the cell; it is negative in sign and often of small enough effect to be ignored.

The sum of the three terms is a negative number except in maximally turgid cells, when it becomes zero.

A Method of Measuring Water Potential (Ψ) of a Plant Tissue

According to one common method of measuring water potential in plant tissues, uniform sample pieces of tissue are placed in a series of solutions of a nonelectrolyte like sucrose or mannitol. The object is to find that solution in which the weight and volume of the tissue does not change, indicating neither a net loss nor a net gain in water. Such a situation would mean that the tissue and the solution were in osmotic equilibrium to begin with, and so the Ψ of the tissues must equal the Ψ of the external solution. Thus, if one can calculate the Ψ of the external solution in which no change in weight or volume of the tissue occurred, one can calculate the Ψ of the tissue.

---

[1]Some authorities (e.g. Kramer et al., 1966) believe that Ψ is determined by Ψ_p, Ψ_π and Ψ_m all acting additively, i.e. $\Psi = \Psi_\pi + \Psi_p + \Psi_m$. However, others (e.g. Salisbury and Ross, 1978) believe that the solution and the colloidal material in a given system or cell may constitute two different phases in equilibrium with each other, hence that a more accurate expression is: $\Psi = \Psi_p + (\Psi_\pi = \Psi_m)$.

Materials and Methods

Prepare 12 beakers (150- or 250-ml) each containing 100 ml of one of the following: distilled water, 0.05, 0.10, 0.15, 0.20, 0.25, 0.30, 0.35, 0.40, 0.45, 0.50 and 0.60 m (molal) sucrose solutions.

The next operation must be done as quickly as possible. Using a cork borer of approximately 1 cm diameter, obtain from a single potato tuber 12 cylinders, each at least 3 cm, and preferably 4 cm, long. Cut all 12 sections to measured <u>uniform length</u> with a razor blade, leaving a clean transverse cut at the end of each cylinder. Place the cylinders between the folds of a moist paper towel, on which positions of the cylinders are denoted by the series of concentrations of sucrose to be used. Using an analytical balance, weigh each cylinder to the nearest milligram. Immediately after each cylinder is weighed, cut the cylinder into uniform slices approximately 2 mm thick, and place the collective slices from one cylinder in one of the test solutions. Do this for each cylinder, being sure that the initial weight of the cylinder placed in each test solution is accurately recorded.

After 1.5-2.0 hours incubation, remove the slices comprising each sample, blot gently on paper towels, and weigh. Repeat this procedure until all samples have been weighed, in the chronological order in which they were initially placed in the test solutions.

Present the data in a table showing original weight, final weight, change in weight, and percentage change in weight, where:

$$\% \text{ Change in weight} = \frac{\text{final weight} - \text{original weight}}{\text{original weight}}$$

Then construct a graph plotting change in weight or % change in weight (on ordinate) versus sucrose concentration (in molality, m) and osmotic potential (in bars) (on abscissa). Calibrate the osmotic-potential axis after first calculating the osmotic potential (Ψ_π) for each sucrose solution. Use the following formula:

$$-\Psi_\pi = miRT$$

where:

m = molality of the solution.
i = ionization constant; numerical value of 1 for sucrose.
R = gas constant (0.083 liter bars/mole degree).
T = absolute temperature (= °C + 273).

Actually, the above formula need be used to calculate the Ψ_π for only one of the sucrose solutions; the Ψ_π values for all other solutions can be more simply calculated using the formula:

$$\frac{m_1}{\Psi_{\pi 1}} = \frac{m_2}{\Psi_{\pi 2}}$$

Determine by interpolation from the graph the sucrose concentration in which no net change in weight occurred. Calculate the Ψ_π for this solution; this value then equals the water potential (Ψ) of the tissue.

References

1. Briggs, G. E. 1967. Movement of Water in Plants. Davis Publishing Company, Philadelphia.

2. Dainty, J. 1965. Osmotic flow. Symposia of the Society for Experimental Biology 19: 75-85.

3. Green, P. B. and F. W. Stanton. 1967. Turgor pressure: direct manometric measurement in single cells of <u>Nitella</u>. Science 155: 1675-1676.

4. Kozlowski, T. T. 1964. Water Metabolism in Plants. Harper and Row, Publishers, New York.

5. Kozlowski, T. T., Ed. 1968a. Water Deficits and Plant Growth. Vol. I. Development, Control, and Measurement. Academic Press, New York.

6. Kozlowski, T. T., Ed. 1968b. Water Deficits and Plant Growth. Vol. II. Plant Water Consumption and Response. Academic Press, New York.

7. Kramer, P. J. 1969. Plant and Soil Water Relationships: A Modern Synthesis. McGraw-Hill Book Company, New York.

8. Kramer, P. J., E. B. Knipling, and L. N. Miller. 1966. Terminology of cell-water relations. Science 153: 889-890.

9. Meyer, B. S., D. B. Anderson, and R. H. Böhning. Introduction to Plant Physiology. D. Van Nostrand Company, New York.

10. Philip, J. R. 1966. Plant-water relations: some physical aspects. Ann. Rev. Plant Physiol. 17: 245-268.

11. Ray, P. M. 1960. On the theory of osmotic water movement. Plant Physiol. 35: 783-795.

12. Ray, P. M. 1972. The Living Plant. 2nd Ed. Holt, Rinehart and Winston, New York.

13. Salisbury, F. B. and C. W. Ross. 1978. Plant Physiology. 2nd Ed. Wadsworth Publishing Company, Belmont, California.

14. Slatyer, R. O. 1967. Plant-Water Relationships. Academic Press, New York.

15. Waring, R. H. and B. D. Cleary. 1967. Plant moisture stress: evaluation by pressure bomb. Science 155: 1248-1254.

16. Wiebe, H. H. 1966. Matric potential of several plant tissues and biocolloids. Plant Physiol. 41: 1439-1442.

17. Wilson, J. W. 1967. The components of leaf water potential. Australian J. Biol. Sci. 20: 329-347; 349-357; 359-367.

Special Materials and Equipment Required
Per Team of 3-4 Students

(1) Large potato (Solanum tuberosum) tuber (local market)
(100 ml each) Following solutions of sucrose: 0.05, 0.10, 0.15, 0.20, 0.25, 0.30, 0.35, 0.40, 0.45, 0.50 and 0.60 molal
(1) Cork borer approximately 1 cm in diameter
(1) Analytical balance

Recommendations for Scheduling

It is recommended that students work in teams of 3 to 4 members, each team performing the entire experiment in a single laboratory period. Sucrose solutions should be prepared in advance.

Duration: 1 laboratory period; approximately 30 minutes at both the beginning and the end of period required for this experiment.

EXERCISE 2

MEASUREMENT OF THE WATER POTENTIAL OF PLANT TISSUES

REPORT

Name _____ Section _____ Date _____

Changes in weight of tissue samples:

Sucrose conc. (m)	Initial wt. (g)	Final wt. (g)	Change in wt. (g)	% Change in wt.
0.00	_____	_____	_____	_____
0.05	_____	_____	_____	_____
0.10	_____	_____	_____	_____
0.15	_____	_____	_____	_____
0.20	_____	_____	_____	_____
0.25	_____	_____	_____	_____
0.30	_____	_____	_____	_____
0.35	_____	_____	_____	_____
0.40	_____	_____	_____	_____
0.45	_____	_____	_____	_____
0.50	_____	_____	_____	_____
0.60	_____	_____	_____	_____

EXERCISE 2

MEASUREMENT OF THE WATER POTENTIAL OF PLANT TISSUES

REPORT

Name _____   Section _____   Date _____

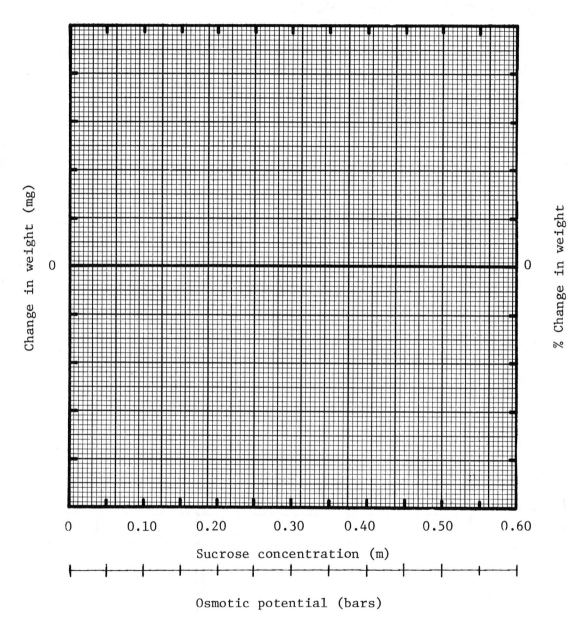

Change in weight (mg)

% Change in weight

0

0

0   0.10   0.20   0.30   0.40   0.50   0.60

Sucrose concentration (m)

Osmotic potential (bars)

Calculation of Ψ of potato tissue:

Exercise 2

Measurement of the Water Potential of Plant Tissues

REPORT

Name _____ Section _____ Date _____

Questions

1. In this experiment the concentration of a solution of sucrose in which no net change in weight (due to gain or loss of water) occurred was determined, and it was stated that the Ψ of the tissue equaled the Ψ_{π} of that solution. What can be said about the comparative Ψ of the tissue and the Ψ of the solution in which no net change in tissue weight occurred? What is the one essential criterion, or the single requisite condition, for osmotic equilibrium on two sides of a differentially permeable membrane, or in all parts of any osmotic system?

2. Is the Ψ of a single cell, tissue or whole plant a constant value? Explain.

3. What can be said about the signs (positive or negative) and relative magnitudes of Ψ_{π} and Ψ_{p} of the potato tissue that was used in this experiment?

EXERCISE 2

MEASUREMENT OF THE WATER POTENTIAL OF PLANT TISSUES

REPORT

Name _____ Section _____ Date _____

4. Describe two other methods for measuring water potentials of plant tissues.

5. Describe methods for measuring Ψ_π and Ψ_p of plant cells and tissues.

6. Briefly describe the relationships among the terms Ψ, Ψ_p and Ψ_π and DPD, TP and OP, and state briefly why the concept and terminology of water potential is preferred.

FACTORS INFLUENCING ENZYME ACTIVITY

Introduction

The purpose of this series of three experiments is to investigate the effects of pH, substrate concentration and enzyme concentration on the rate of an enzyme-catalyzed reaction. The reaction selected for investigation is the decomposition of hydrogen peroxide to water and oxygen, which is catalyzed by the enzyme catalase:

$$H_2O_2 \xrightarrow{\text{Catalase}} H_2O + \tfrac{1}{2}O_2$$

Activity will be assayed in crude enzyme extracts prepared from spinach leaves by a macro-manometric procedure.

Catalase has been selected for these experiments because it is ubiquitous in aerobic cells, it is a relatively stable enzyme, and the primary reaction which it catalyzes can be conveniently measured by collecting the oxygen which is produced.

Materials and Methods

The class will work in three (3) groups, each group performing a different experiment. However, every student will be expected to report on all three experiments. Each group set up an apparatus as illustrated in Fig. 3-1. Prepare a crude enzyme extract by grinding approximately 12 g of fresh spinach leaf tissues in 100 ml of distilled water, using a mortar and pestle. Add a small amount of acid-washed, ignited sand to expedite the homogenization. Allow the large particulate material to settle out of the homogenate, and use an agar (large-bore) pipet to transfer the crude enzyme extract. Make a trial run using 20 ml of pH 7.2 buffer (see Table 3-I), 2.0 ml of enzyme extract and 10 ml of 3% H_2O_2. Mix the enzyme extract and buffer in the reaction vessel; pour the H_2O_2 into the separatory funnel (with stopcock closed). Mount the reaction vessel assembly on a ring stand with a clamp, and submerse the reaction vessel in a constant temperature water bath set at 30°C. Connect the reaction vessel assembly to an inverted 50-ml buret with the rubber and glass tubing provided. Fill the inverted buret with water by suction and close off. After allowing approximately 5 minutes for thermal equilibration between the contents of the reaction vessel and the water bath, simultaneously open the stopcock on the separatory funnel (allowing the H_2O_2 to run into the reaction vessel) and the pressure release. Mark the time when all the H_2O_2 has run into the reaction vessel. <u>Close the stopcock on the separatory funnel and the pressure release</u>. Agitate the reaction vessel gently and continuously. Note the volume of O_2 collected in the buret during the succeeding 5 minutes, or the time required to collect 50 ml of O_2 if less than 5 minutes. A desirable reaction rate is approximately 10 ml O_2/minute. If the activity observed in the trial run exceeds this rate, dilute the enzyme extract with distilled water before

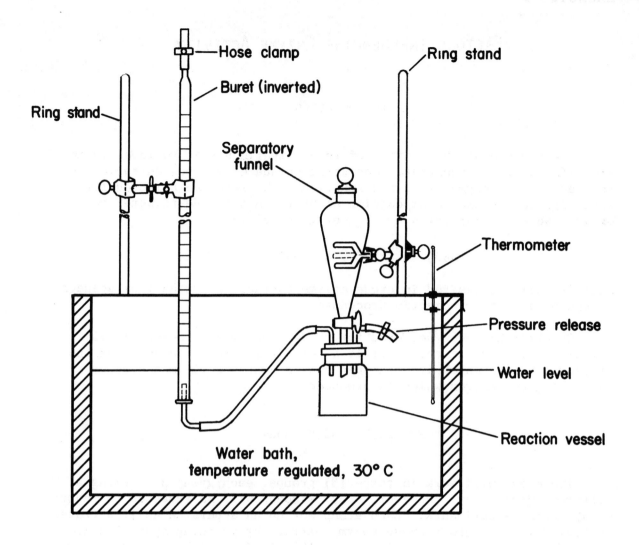

Fig. 3-1. Apparatus for assaying catalase activity.

Table 3-I.   Preparation of Buffers

pH	Stock solutions		Amounts (ml)		Total volume (ml)
	A	B	A	B	
4.0	0.1 M Citric acid ($C_6H_8O_7 \cdot H_2O$)	0.2 M Dibasic sodium phosphate (Na_2HPO_4)	61.4	38.6	200
5.0	0.1 M Citric acid ($C_6H_8O_7 \cdot H_2O$)	0.2 M Dibasic sodium phosphate (Na_2HPO_4)	48.6	51.4	200
5.4	0.1 M Citric acid ($C_6H_8O_7 \cdot H_2O$)	0.2 M Dibasic sodium phosphate (Na_2HPO_4)	44.4	55.6	200
6.0	0.1 M Citric acid ($C_6H_8O_7 \cdot H_2O$)	0.2 M Dibasic sodium phosphate (Na_2HPO_4)	35.8	64.2	200
6.6	0.2 M Monobasic sodium phosphate (NaH_2PO_4)	0.2 M Dibasic sodium phosphate (Na_2HPO_4)	62.5	37.5	200
7.2	0.2 M Monobasic sodium phosphate (NaH_2PO_4)	0.2 M Dibasic sodium phosphate (Na_2HPO_4)	28.0	72.0	200
7.8	0.2 M Monobasic sodium phosphate (NaH_2PO_4)	0.2 M Dibasic sodium phosphate (Na_2HPO_4)	8.5	91.5	200
8.4	0.2 M KCl+0.2M H_3BO_3	0.2 M NaOH	50.0	8.6	200
9.0	0.2 M KCl+0.2M H_3BO_3	0.2 M NaOH	50.0	21.4	200
10.0	0.2 M KCl+0.2M H_3BO_3	0.2 M NaOH	50.0	39.1	200
12.0	0.2 M KCl+0.2M H_3BO_3	0.2 M NaOH	50.0	61.0	200

proceeding with the assigned experiment. Select a standard time interval (3 to 5 minutes) to be used in all measurements. For all experiments, immediately after the final enzyme extract is prepared pipet aliquots of extract into test tubes containing the appropriate buffer, set the test tubes in a rack, and set the rack of tubes in the constant-temperature water bath.

Experiment 1. Effect of pH on the rate of the reaction. Follow the general procedure described for making the trial run, using as standard conditions: temperature of 30° C; 20 ml of buffer, 2 ml of enzyme extract and 10 ml of 3% H_2O_2 in each reaction mixture; time interval of 3 to 5 minutes. Use buffer at each of the following pH's (see Table 3-I): 4.0, 5.0, 5.4, 6.0, 6.6, 7.2, 7.8, 8.4, 9.0, 10.0 and 12. Repeat the experiment if time permits. Record for each trial the reaction rate (ml O_2 collected/min). Plot a graph of reaction rate versus pH.

Experiment 2. Effect of enzyme concentration on the rate of the reaction. Prepare a crude enzyme extract and make a trial run as described above. Use buffer at pH 7.2 (see Table 3-I) and a temperature of 30° C exclusively. Determine the reaction rate with each of the following concentrations of enzyme. (Each reaction mixture should contain 20 ml of buffer, 2 ml of enzyme and 10 ml of substrate.)

0.00 ml 0.25 0.50 1.00	Make up to a volume of 2.0 ml with buffer.
2.00	Use 20 ml of buffer, 10 ml of 3% H_2O_2.
3.00	Use 19 ml of buffer, 10 ml of 3% H_2O_2.
4.00	Use 18 ml of buffer, 10 ml of 3% H_2O_2.

Record for each trial the reaction rate (ml O_2 collected/min). Repeat the experiment if time permits. Plot a graph of reaction rate versus enzyme concentration.

Experiment 3. Effect of substrate concentration on the rate of the reaction, using two standard enzyme concentrations. Prepare an enzyme extract and make a trial run as described previously. Use buffer at pH 7.2 and 30° C exclusively. Remove 20 ml of enzyme extract and dilute it with 20 ml of buffer; call this preparation [E]/2. Retain the remaining enzyme extract and designate its concentration as [E]. Use each of the following quantities of 3% H_2O_2 first with 2.0 ml of [E]/2 solution; then repeat with 2.0 ml of [E] solution. Each reaction mixture should contain 20 ml of buffer, 2 ml of enzyme solution and 10 ml of substrate.

Volume of 3% H_2O_2

0.0 ml 0.5 1.0 2.0 4.0 8.0	Add sufficient buffer to the separatory funnel to make a total volume of substrate equal to 10 ml.
10.0	

If it is necessary to increase the amount of H_2O_2 further to saturate the enzyme, use 2.0 ml of 30% H_2O_2 plus 8.0 ml of buffer (this is

26

equivalent to 20 ml of 3% H_2O_2). Record the reaction rates. Plot two graphs (one for [E] and one for [E]/2) on the same set of coordinates.

References

1. Bernhard, S. 1968. The Structure and Function of Enzymes. W. A. Benjamin, New York.

2. Bonner, J. and J. E. Varner, Eds. 1976. Plant Biochemistry. 3rd Ed. Academic Press, New York.

3. Boyer, P. D., H. Lardy, and K. Myrbäck, Eds. 1959-1963. The Enzymes. 8 volumes. Academic Press, New York.

4. Brandts, J. F. 1967. Heat effects on proteins and enzymes. In: A. H. Rose, Ed. Thermobiology. Academic Press, New York.

5. Cleland, W. W. 1967. Enzyme kinetics. Ann. Rev. Biochem. 36: 77-112.

6. Colowick, S. P. and N. D. Kaplan, Eds.-in-Chief. 1955-1972. Methods in Enzymology. 25 volumes. Academic Press, New York.

7. Dixon, M. and E. C. Webb. 1964. Enzymes. Academic Press, New York.

8. Giese, A. C. 1968. Cell Physiology. 3rd Ed. W. B. Saunders Company, Philadelphia.

9. Lehninger, A. L. 1975. Biochemistry. 2nd Ed. Worth Publishers, New York.

10. Mahler, H. R. and E. H. Cordes. 1971. Biological Chemistry. 2nd Ed. Harper and Row, Publishers, New York.

11. Salisbury, F. B. and C. W. Ross. 1978. Plant Physiology. 2nd Ed. Wadsworth Publishing Company, Belmont, California.

Special Materials and Equipment Required
Per Team of 3-8 Students

(Approximately 25 g) Fresh spinach (Spinacia oleracea) leaves (local market)
(Approximately 10 g) Acid-washed, ignited sand
(1) Mortar and pestle
(1) Enzyme assay apparatus (see Fig. 3-1):
        (1) 50-ml buret
        (1) 60-ml separatory funnel
        (1) Reaction vessel (e.g. wide-mouth specimen bottle, 100-ml)
        (1) Rubber stopper to fit reaction vessel

(Approximately 36 cm) Rubber tubing, 3/16- or 1/4-inch I.D.
(2 pieces) Glass tubing, 7 mm O.D., bent as illustrated in
        Fig. 3-1.
        (2) Clamps, Hoffman or Mohr type
(Approximately 500 ml) 3% H_2O_2, freshly prepared from 30% H_2O_2
(250 ml) Each of the following buffer stock solutions:
        0.2 M Na_2HPO_4 (Experiments 1, 2 and 3)
        0.1 M Citric acid (Experiment 1)
        0.2 M NaH_2PO_4 (Experiments 1, 2 and 3)
        0.2 M KCl + 0.2 M H_3BO_3 (Experiment 1)
        0.2 M NaOH (Experiment 1)
(300 to 600 ml) Phosphate buffer, pH 7.2 (Experiments 2 and 3); approxi-
        mately 100 ml of each buffer needed for Experiment 1, except
        250 ml of pH 7.2 buffer.
(1) 2- to 10-ml agar (large bore) pipet (Experiments 1 and 2)
(1 each) 1-ml and 10-ml agar pipets (Experiment 3)
(1) Thermoregulated water bath

Recommendations for Scheduling

    It is suggested that the students in each laboratory section work in
three teams, each team performing one complete experiment. Each student
should report on all three experiments. The buffers (at least the buffer
stock solutions) and 3% H_2O_2 should be prepared and the essential components of the assay apparatus should be prepared in advance of the start
of the laboratory period.

Duration: 1 laboratory period; exclusive attention to this experiment for entire period required.

EXERCISE 3

FACTORS INFLUENCING ENZYME ACTIVITY

REPORT

Name _____   Section _____   Date _____

Experiment 1 – Effect of pH:

pH	ml O_2 collected/minute		
	Trial 1	Trial 2	Mean
4.0	_____	_____	_____
5.0	_____	_____	_____
5.4	_____	_____	_____
6.0	_____	_____	_____
6.6	_____	_____	_____
7.2	_____	_____	_____
7.8	_____	_____	_____
8.4	_____	_____	_____
9.0	_____	_____	_____
10.0	_____	_____	_____
12.0	_____	_____	_____

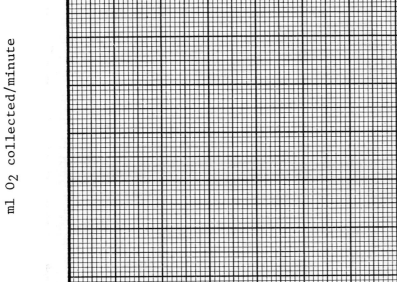

ml O_2 collected/minute

pH

4   5   6   7   8   9   10   11   12

Exercise 3

Factors Influencing Enzyme Activity

REPORT

Name _____ Section _____ Date _____

Experiment 2 - Effect of Enzyme Concentration:

Enzyme conc. (ml)	ml O_2 collected/minute		
	Trial 1	Trial 2	Mean
0.00	_____	_____	_____
0.25	_____	_____	_____
0.50	_____	_____	_____
1.00	_____	_____	_____
2.00	_____	_____	_____
3.00	_____	_____	_____
4.00	_____	_____	_____

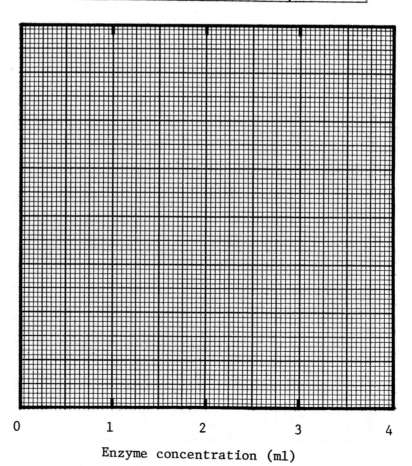

ml O_2 collected/minute

Enzyme concentration (ml)

EXERCISE 3

FACTORS INFLUENCING ENZYME ACTIVITY

REPORT

Name _____   Section _____   Date _____

Experiment 3 - Effect of Substrate Concentration:

[E]/2				[E]			
Substrate conc. (ml 3% H_2O_2)	ml O_2 collected/minute			Substrate conc. (ml 3% H_2O_2)	ml O_2 collected/minute		
	Trial 1	Trial 2	Mean		Trial 1	Trial 2	Mean
0.0	_____	_____	_____	0.0	_____	_____	_____
0.5	_____	_____	_____	0.5	_____	_____	_____
1.0	_____	_____	_____	1.0	_____	_____	_____
2.0	_____	_____	_____	2.0	_____	_____	_____
4.0	_____	_____	_____	4.0	_____	_____	_____
8.0	_____	_____	_____	8.0	_____	_____	_____
10.0	_____	_____	_____	10.0	_____	_____	_____

ml O_2 collected/minute

Substrate concentration (ml 3% H_2O_2)

0   1   2   3   4   5   6   7   8   9   10

EXERCISE 3

FACTORS INFLUENCING ENZYME ACTIVITY

REPORT

Name _____ Section _____ Date _____

Questions

1. Write a concise explanation of the results of each of the three experiments.

2. Explain the mechanism by which catalase accelerates the rate of decomposition of hydrogen peroxide, in terms of the concept originally proposed by L. Michaelis and M. L. Menten.

3. In what ways does hydrogen-ion concentration affect enzyme activity?

FACTORS INFLUENCING ENZYME ACTIVITY

REPORT

Name _____   Section _____   Date _____

4. By analysis of the curves plotted for Experiment 3, is the apparent
    Michaelis-Menten constant (K_m) for hydrogen peroxide evidently the
    same for the two enzyme concentrations used? Explain.

5. Give some indication of the significance of catalase in cellular
    metabolism by describing two or three specific reactions, common
    to plant cells, in which hydrogen peroxide is formed.

6. With reference to question 2, cite specific data indicating how much
    the energy of activation for the decomposition of hydrogen peroxide
    is reduced in the presence of catalase, compared to the non-enzymic
    thermochemical reaction.

THIN-LAYER CHROMATOGRAPHY OF CHLOROPLAST PIGMENTS AND DETERMINATION OF PIGMENT ABSORPTION SPECTRA

Introduction

Contained in the chloroplasts of all higher plants are two major kinds of pigments which absorb light energy effective in photosynthesis. These are the green chlorophylls (a and b) and the commonly red, orange or yellow carotenoids. The carotenoids, several types of which typically occur in a single plant, are of two fundamental types, the carotenes and the oxygenated carotenoids or xanthophylls. Structural formulas of chlorophylls a and b; β-carotene, a representative carotene; and lutein, a representative xanthophyll, are illustrated in Figure 4-1.

Chlorophylls a and b possess a porphyrin structure, comprised of four pyrrole rings, with a chelated magnesium atom at its center, a fifth 5-carbon ring, and a 20-carbon phytyl group attached to one of the pyrrole rings. The empirical formulas of chlorophylls a and b are $C_{55}H_{72}O_5N_4Mg$ and $C_{55}H_{70}O_6N_4Mg$, respectively. The difference lies in the substituent on ring 3. Chlorophyll a has a methyl ($-CH_3$) group in that position; chlorophyll b has an aldehyde ($-CHO$) group. In most higher plants and green algae chlorophylls a and b occur in a ratio of 2-3 : 1.

Carotenoids are 40-carbon compounds of which a large number of distinct types are known. The carotenes are pure hydrocarbons; β-carotene, for example, has the empirical formula $C_{40}H_{56}$. Xanthophylls contain oxygen in the terminal rings, and many, but not all, have the empirical formula $C_{40}H_{56}O_2$. Carotenoids are not invariably red, orange or yellow. Some are green, others pink, and some quite black. Neither are carotenoids confined in occurrence to plants; indeed they occur in all major phyla of plants and animals. In higher plants they are not confined to chloroplasts but often are found in other types of plastids, called chromoplasts, in fruits, flower parts, etc.

Chlorophylls and carotenoids occur in close association with each other and with the protein and lipid membrane constituents in the grana of chloroplasts. Whereas it was formerly believed that the photosynthetic pigments occur exclusively in grana lamellae, recent evidence indicates that chlorophyll at least occurs also in the stroma lamellae.

All photosynthetic plants, except photosynthetic bacteria, possess chlorophyll a, but many do not contain the combination of chlorophylls a and b, which is characteristic of green algae and higher plants. Photosynthetic bacteria possess unique forms of chlorophyll. Blue-green algae possess only chlorophyll a. Brown algae and diatoms possess chlorophylls a and c. Red algae commonly have chlorophylls a and d. The red and blue-green algae also possess phycobilin pigments, which, while not occurring in plastids, nevertheless absorb light energy which is effectively transferred to chlorophyll and therefore serve as photosynthetic pigments. All the pigments discussed above, other than chlorophyll a, often are termed accessory pigments.

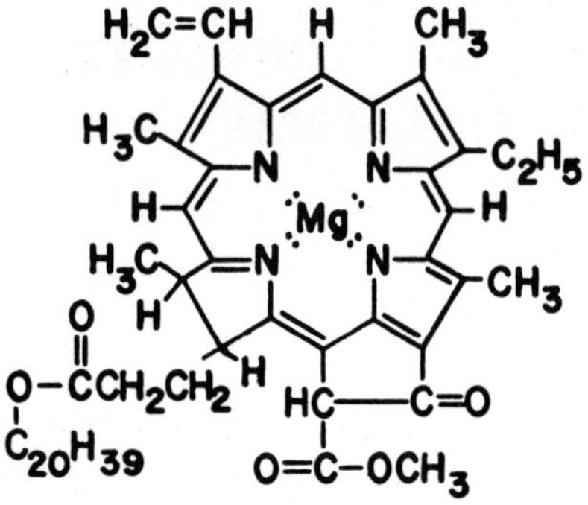

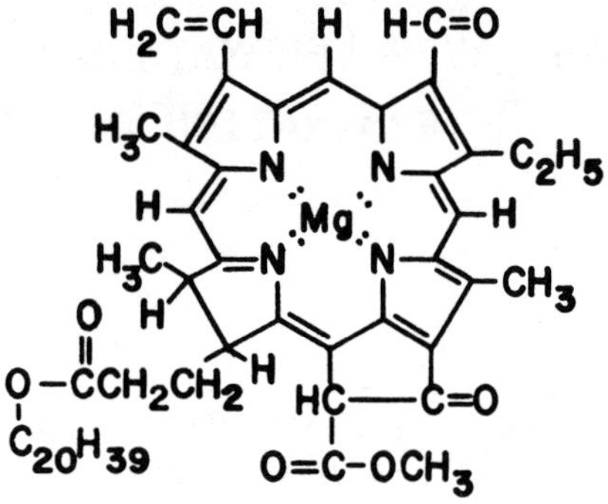

Chlorophyll a

Chlorophyll b

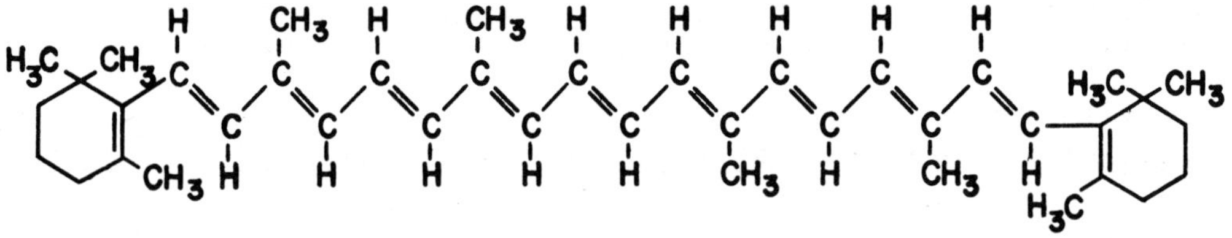

β – Carotene

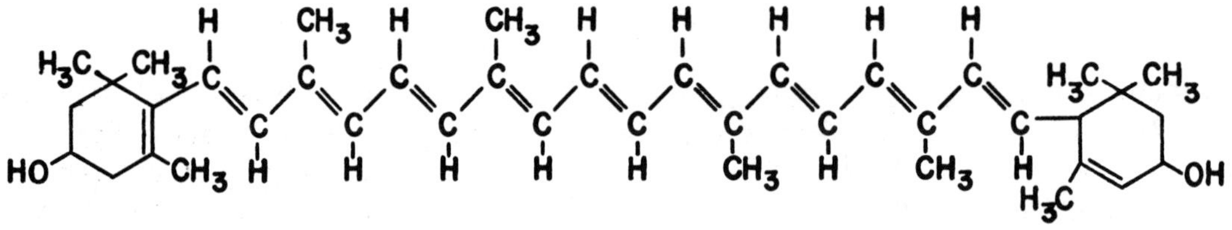

Lutein

Fig. 4-1. Structures of four common chloroplast pigments (data of R. C. Dougherty et al., 1966, J. Amer. Chem. Soc. 88:5037 in Plant Physiology by Frank B. Salisbury and Cleon Ross, 1969, Wadsworth Publishing Company, Inc., Belmont, California 94002).

The purpose of this exercise is to extract the chloroplast pigments from leaves of spinach, separate the pigments by thin-layer chromatography, determine the absorption spectra of certain of the pigments, and to identify certain of the pigments by their chromatographic behavior and absorption spectra. Before undertaking the procedure, it will be well to become familiar with the technique of thin-layer chromatography.

Thin-layer chromatography (TLC) is a relatively new method of adsorption and partition chromatography on a micro scale. It supplements the previously known methods of column, paper, ion exchange and precipitation chromatography. In operation, a glass plate or other support is coated with a thin layer of adsorbent which adheres to the plate. The layer represents an "open" column. The sample is spotted on this coating as a solution and is developed in a tank containing a small amount of the appropriate solvent, normally by the ascending technique. Only short travel distances (approximately 10-15 cm) are required.

The TLC method was given major impetus by Stahl, who in 1956, described not only the first practical equipment for coating glass plates on a standardized basis, but who also expanded the horizons of TLC by applying it to many problems in diversified fields of interest. The TLC method continues to be improved, and new applications of the method are being found continually. Among the most notable recent developments are: (a) commercial production of a large variety of adsorbents, pre-coated and quite durable glass plates, pre-coated thin films, etc.; (b) commercial production of precision coating apparatus; and (c) adaptation of radio-chromatogram scanning instruments to thin-layer chromatograms.

Recent developments in thin-layer chromatography have demonstrated conclusively that the technique has not only closed an existing gap as a micro-chromatographic method but that it is also finding wide application in new areas and replacing paper techniques in others. This situation has resulted from the significant advantages of TLC, which include:

1. Much faster development than column or paper chromatography. The separating time, depending on the solvent used and layer thickness, ranges from 20-40 minutes for most applications.

2. Pure inorganic adsorption layers which:
   a. permit the use of extremely harsh (corrosive) spray reagents and which
   b. do not exhibit undesirable integral fluorescence.

3. Noticeably sharper separations than can be obtained with paper or conventional column chromatography.

4. A higher sensitivity. Normally, sensitivity of TLC is from 10 up to 100 times greater than that of paper chromatography.

5. Requires only limited sample quantities with a minimum of approximately 0.5 μg, yet up to 500 μg can be applied to one spot. It is an ultra micro method suited to trace analysis.

6. Results may be correlated to column separations because, in general, the same materials are used as adsorbents.

Basically, by selecting a suitable adsorption climate, adsorbent and solvent, it is possible to separate any compound if there is a difference in the adsorption properties of the fraction involved. The fractions will separate in a distinct sequence, and the following "rules of thumb" are based on previously known facts.

1. Saturated hydrocarbons are not absorbed at all, or only slightly.

2. The adsorption of unsaturated hydrocarbons increases with the number of double bonds and with the number which are in conjugation.

3. If functional groups are introduced into a hydrocarbon, the adsorption affinity increases in the following sequence: $-CH_3$, $-O-Alkyl$, $>C=O$, $-NH_2$, $-OH$, $-COOH$

Materials and Methods[1]

Prepare a slurry of silica gel G by vigorously shaking 30 g of the silica gel with 60 ml of distilled water in a stoppered 250- or 500-ml Erlenmeyer flask. The slurry must be prepared quickly and used immediately, because the silica gel contains $CaSO_4$, which acts as a binder. Using the special mounting board and applicator provided, coat a number of 5 x 20 and 20 x 20 cm glass plates with the slurry of silica gel G. The applicator should be set to apply a layer of adsorbent 250 μ thick. Place the coated plates in a drying rack and store them in a desiccator until time to apply the samples to be chromatographed.

Extract chloroplast pigments from spinach leaves by grinding approximately 25 g of fresh leaves (or approximately 2.5 g of dried leaves) in approximately 100 ml of acetone (or 80% acetone if dried leaves are used) in an homogenizer. When the acetone is colored deep green, add several grams of Na_2SO_4 to remove water from the extract, and filter the preparation on a Büchner funnel. Evaporate the acetone extract in a rotary-film evaporator, and re-dissolve the dried pigments in approximately 10 ml of chloroform.

Select a 5 x 20 cm plate coated with adsorbent and run a sheet of tissue around the margins to remove excess or flaking gel. On the end of the plate to which the sample is to be applied make small scratches on both sides at a distance of 1.5 cm from the end. Then scratch a small mark at a distance of 15 cm on each side of the plate above the original marks. A special labeling template may be provided to expedite the

---

[1]The assistance of Dr. Norman I. Bishop, Department of Botany and Plant Pathology, Oregon State University, Corvallis, in the design of these procedures is gratefully acknowledged.

marking of the plates. The two sets of marks described above mark the origin and the position to which the solvent will be allowed to migrate, respectively.

Next, using a 0.01 or 0.1 ml pipet or a special micrometer buret, apply a small aliquot of the extract between the two marks at the origin end of the plate. Be sure the sample is spotted near the center of the origin end of the plate. Avoid touching the buret or pipet to the adsorbent, and keep the sample spot 1 cm or smaller in diameter. A cool-air fan may be used to hasten evaporation of the solvent if repeated application of dilute extract is necessary.

When the sample spot is thoroughly dry, place the plate in a developing chamber containing solvent (hexane-diethyl ether-acetone, 60:30:20, v/v/v). The depth of the solvent is critical, since the sample spot must not dip into the solvent. Observe the development process and the differential migration of the various pigments up the chromatogram. When the solvent front has advanced 15 cm above the origin, remove the chromatogram from the developing tank and allow it to dry.

Make a scaled diagram of the chromatogram, showing all of the pigments which are visible. With the assistance of the instructor, identify as many of the pigments as possible (see reference 4 for assistance with pigment identification). Calculate the R_f value for each of the pigments. Add the identifications and R_f values to the diagram.

If time permits, prepare a two-dimensional chromatogram of the chloroplast pigments, using a 20 x 20 cm glass plate and following the instructions given in the laboratory. Record the results in a fully-labeled diagram drawn to scale.

As soon as inspection of the chromatograms has been completed and all data recorded, remove bands of silica gel containing the individual pigments using a clean razor blade. Immediately place each sample of gel containing a single pigment in a centrifuge tube and add approximately 5 ml of acetone. Agitate the tube to facilitate elution of the pigment off the gel. Centrifuge the tubes to sediment the gel. Then pipet off the supernatant from each tube, and transfer the pigment solution to a labeled, stoppered test tube and place in the refrigerator until further use.[2]

In the next laboratory period, adjust the concentration of each pigment solution by dilution or condensation, as is found to be necessary, and determine the absorption spectra of a number of the pigments, including chlorophyll a, chlorophyll b, β-carotene, and neoxanthin or lutein using a Beckman DB spectrophotometer and recorder, a Bausch and Lomb Spectronic 20 colorimeter-spectrophotometer, or other suitable instrument. Measure the

---

[2]To minimize degradation, the pigments should be eluted as soon as possible, and the solutions should be stored in darkness at low temperature. Covering of the dry chromatogram with a glass plate and taping the edges airtight is advisable if the pigments cannot be eluted within 10 to 15 minutes following drying.

absorbance [or optical density (O.D.)] or % absorption of each pigment solution over the range of wave lengths from 400 to 700 nm. If a non-recording spectrophotometer is used, measure the absorbance or % absorption manually at 10-nm intervals. Plot the absorption spectra, and include the graphs in your report.

References

1. Frey-Wyssling, A. and K. Mühlethaler. Ultrastructural Plant Cytology, with an Introduction to Molecular Biology. Elsevier Publishing Company, Amsterdam.

2. Goodwin, T. W. 1965. Chemistry and Biochemistry of Plant Pigments. Academic Press, New York.

3. Goodwin, T. W., Ed. 1966-1967. Biochemistry of Chloroplasts. 2 volumes. Academic Press, New York.

4. Hirayama, O. 1967. Lipids and lipoprotein complex in photo-synthetic tissues. II. Pigments and lipids in blue-green alga, Anacystis nidulans. J. Biochem. 61: 179-185.

5. Jensen, W. A. 1970. The Plant Cell. 2nd Ed. Wadsworth Publishing Company, Belmont, California.

6. Jensen, W. A. and R. B. Park. 1967. Cell Ultrastructure. Wadsworth Publishing Company, Belmont, California.

7. Kirk, J. T. O. and R. A. E. Tilney-Bassett. 1967. The Plastids. Their Chemistry, Structure, Growth and Inheritance. W. H. Freeman and Company, London.

8. Menke, W. 1962. Structure and chemistry of plastids. Ann. Rev. Plant Physiol. 13: 27-44.

9. Ò hEocha, C. 1965. Biliproteins of algae. Ann. Rev. Plant Physiol. 16: 415-434.

10. Randerath, K. 1966. Thin-layer Chromatography. 2nd Ed. Academic Press, New York.

11. Salisbury, F. B. and C. W. Ross. 1978 Plant Physiology. 2nd Ed. Wadsworth Publishing Company, Belmont, California.

12. Smith, J. H. C. and C. S. French. 1963. The major and accessory pigments in photosynthesis. Ann. Rev. Plant Physiol. 14: 181-224.

13. Stahl, E. 1969. Thin-layer Chromatography. 2nd Ed. Tr. by M. R. F. Ashworth. Springer-Verlag, Berlin, Heidelberg, New York.

14. Vernon, L. P. and G. R. Seely, Eds. 1966. The Chlorophylls. Academic Press, New York.

15. Willard, H. H., L. L. Merritt, Jr., and J. A. Dean. 1965. Instrumental Methods of Analysis. 4th Ed. D. Van Nostrand Company, Princeton, New Jersey.

Special Materials and Equipment Required
Per Team of 2 Students

*(Approximately 25 g fresh or 2.5 g dried) Spinach (Spinacia oleracea) leaves (local market)
*(Approximately 100 ml) Acetone
*(10 ml) Chloroform
*(1) Rotary-film evaporator and flask
*(1) 5 x 20 cm glass plate pre-coated with a 250 μ layer of silica gel G (Brinkmann Instruments, Incorporated, Cantiague Road, Westbury, New York, 11590, or nearest branch office) and stored in a desiccator
 (1) Developing tank for 5 x 20 cm plates (e.g. Desaga Round Developing Tank, 2 3/8 x 9 inch, available from Brinkmann Instruments, Incorporated)
 (1) 0.01- or 0.1-ml long-tip pipet, capillary pipet or micrometer buret
 (Approximately 25 ml) of a solvent system composed of hexane-diethyl ether-acetone (6:3:2, v/v/v)
*(1) 20 x 20 cm glass plate pre-coated with a 250 μ layer of silica gel G
*(1) Developing tank for 20 x 20 cm plates (e.g. Desaga Rectangular Developing Tank, available from Brinkmann Instruments, Incorporated)
*(1 each) Mounting board, adjustable applicator and desiccator (available, e.g., from Brinkmann Instruments, Incorporated)
 (1) Colorimeter or spectrophotometer

Recommendations for Scheduling

*It is recommended that a single pigment extract be prepared, as a demonstration, for each laboratory section. Thin-layer plates should be prepared at least 12 hours in advance and stored in a desiccator. Alternatively, commercially available pre-coated plates may be used, although the author has had better success with plates prepared in the laboratory. At least one 20 x 20 cm chromatogram should be prepared, and the appropriate pigments subsequently eluted and combined with the collective samples eluted from the 5 x 20 cm chromatograms, to be certain of having an adequate quantity of each pigment for determination of absorption spectra.

The author demonstrates the determination of absorption spectra for two or more pigments on a recording spectrophotometer (Beckman DB Spectrophotometer with recorder) and then has student teams determine the spectra manually on Bausch & Lomb Spectronic 20 Colorimeter-Spectrophotometers.

Duration: 2 laboratory periods; approximately 1.5 to 2 hours exclusive attention to this experiment required each period.

EXERCISE 4

THIN-LAYER CHROMATOGRAPHY OF CHLOROPLAST PIGMENTS AND DETERMINATION OF PIGMENT ABSORPTION SPECTRA

REPORT

Name _____ Section _____ Date _____

Results of chromatography:

Pigment   R$_f$

solvent front

origin

X 3/4

THIN-LAYER CHROMATOGRAPHY OF CHLOROPLAST PIGMENTS AND DETERMINATION OF PIGMENT ABSORPTION SPECTRA

REPORT

Name _____ Section _____ Date _____

Absorption spectra:

EXERCISE 4

THIN-LAYER CHROMATOGRAPHY OF CHLOROPLAST PIGMENTS AND DETERMINATION OF PIGMENT ABSORPTION SPECTRA

REPORT

Name _____ Section _____ Date_____

Questions

1. What wavelength(s) and color(s) of light does each pigment absorb most strongly?

2. How would you determine an action spectrum for photosynthesis?

3. How does an action spectrum for photosynthesis in angiosperms generally compare with the absorption spectra for the chlorophylls, carotenes and xanthophylls which you isolated? What conclusions can you make from a comparison of the absorption spectra and the action spectrum for photosynthesis?

4. Explain the basis for separation of compounds by thin-layer chromatography.

EXERCISE 5

EFFECTS OF TEMPERATURE AND LIGHT INTENSITY ON THE RATE OF PHOTOSYNTHESIS IN A GREEN ALGA

The purpose of this experiment is to measure, by a sensitive manometric method, the rate of photosynthesis in a unicellular green alga as a function of light intensity at two temperatures under conditions where carbon dioxide is in abundant supply. A very large part of the research on photosynthesis has been done with cultures of unicellular green algae, which afford several conveniences compared to terrestrial seed plants. The unicellular algae can be grown in defined inorganic culture media under standard and reproducible conditions; uniform, dense, very active samples can be prepared and pipetted as a liquid suspension; and rate of photosynthesis can be accurately measured as rate of oxygen production by sensitive manometric methods.

In this experiment the rate of photosynthesis in the unicellular green alga Scenedesmus obliquus will be measured as rate of oxygen production, using a manometric method. The rate of photosynthesis will be measured over a range of light intensities at one temperature; then the measurements will be repeated over the same range of light intensities at a 10° C higher temperature. Of course while oxygen production in photosynthesis is occurring, oxygen consumption in aerobic respiration is also occurring. Hence, in order to measure the true photosynthetic rate, it is necessary to correct the apparent or net photosynthetic rate of oxygen production by determining also the rate of simultaneous oxygen consumption in respiration. The true photosynthetic rate is then the sum of the observed net rate of oxygen production and the rate of oxygen consumption in respiration. It is commonly assumed that the rate of mitochondrial respiration of photosynthesizing cells is equal to the rate of mitochondrial respiration of the same cells in darkness, temperature and other factors being equal, although for at least some plants (e.g. Chlorella) this may not be strictly correct.

Materials and Methods[1]

The green alga to be used in this experiment is Scenedesmus obliquus. Suspensions of the alga were grown on an inorganic medium in the presence of air enriched to a 4% carbon dioxide concentration. The composition of the culture medium is as follows:

---

[1]The basic procedures utilized in this experiment were contributed by Dr. Norman I. Bishop, Department of Botany and Plant Pathology, Oregon State University, Corvallis, whose assistance is gratefully acknowledged.

KNO_3	0.809 g
$NaCl$	0.468
Na_2HPO_4	0.178
$NaH_2PO_4 \cdot 2H_2O$	0.468
$CaCl_2$	0.022
$MgSO_4 \cdot 7H_2O$	0.247
$FeSO_4 \cdot 7H_2O$	0.020
$MnCl_2 \cdot 2H_2O$	0.0002
$ZnSO_4 \cdot 7H_2O$	0.0001
Versene	0.020

all dissolved in 1 liter
of glass distilled water.

Routinely, culture medium is autoclaved in 250-ml aliquots. After cooling, aliquots of medium are inoculated with Scenedesmus, which is otherwise maintained on an agar medium in test tubes. Sufficient growth occurs in the liquid cultures within 3 to 5 days to insure an adequate number of cells for photosynthesis experiments.

Before describing the precise procedure to be followed in this experiment, it is necessary to describe briefly the principle and operation of the instrument to be employed. The instrument to be used is a Gilson Respirometer—Refrigerated, Photosynthesis Model. An examination of the instrument will reveal that it consists basically of: (1) a thermoregulated water bath with a transparent bottom; (2) a battery of incandescent lamps mounted below the transparent bottom of the water bath; and (3) 14 manometers, each with a digital micrometer, mounted on a shaking device. In practice, a reaction flask is attached to one arm of each manometer, and the flask is submerged in the constant temperature water bath. To the system is attached a reference flask which permits simultaneous compensation for variations in temperature and barometric pressure in all of the experimental flasks.

The differential manometer (Fig. 5-1) is based on the principle that at constant temperature and constant gas volume any change in the amount of a gas can be measured by a change in its pressure. An increase in pressure on the flask side of the manometric fluid results from oxygen production in photosynthesis, a decrease in pressure from oxygen consumption in respiration. A calibrated digital micrometer connected to each manometer is used to return the manometric fluid to its balanced position, and the microliters of oxygen produced in photosynthesis (or consumed in respiration) is read directly on the micrometer.[2]

---

[2]To convert microliter (μl) micrometer readings to standard conditions, multiply by the following correction factor: $\dfrac{(273)(P_b)}{(t+273)(760)}$

where:
t = water bath temperature in degrees C.
P_b = operating pressure (usually the same as barometric pressure).
760 is standard barometric pressure.
273 converts temperature readings into degrees K.

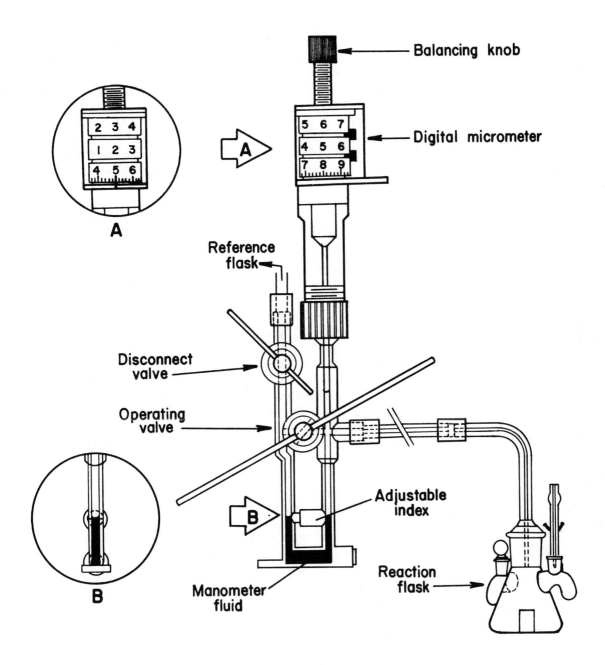

Fig. 5-1. Diagram of a Gilson volumometer (manometer and attached digital micrometer) and attached reaction flask. The operating valve and disconnect valve both are shown in the operating position. Inserts A and B illustrate face-on views of the micrometer and the manometric fluid in open arm of the manometer, respectively.

Set up the actual experiment as follows: Prepare 200 ml of carbonate-bicarbonate buffer at pH 9.08 by mixing 30 ml of 0.1 M Na_2CO_3 and 170 ml of 0.1 M $NaHCO_3$. Centrifuge down approximately 250 ml of the algal suspension provided; then wash the cells with buffer. Centrifuge again, and resuspend the cells in 50 ml of buffer. Pipet 3 ml of dense algal suspension into the main compartment of each of the 14 flasks. Attach the flasks to the manometers, and, with the operating valves open (up), set each digital micrometer at 200 µl. Submerge the flasks in the water bath set at 20°C, and allow 10 to 20 minutes equilibration in darkness. (The equilibration period is required for thermal equilibration of the flask contents with the water bath and for equilibration between the carbon dioxide in solution and in the gas phase in the flasks.) Close the operating valves, and 10 minutes later record the volume of oxygen consumed in each flask. Calculate the mean rate of dark respiration (µl O_2 consumed/10 minutes); this value is to be added to each net photosynthetic rate measured at 20°C when true photosynthetic rates are calculated. Now open the operating valves, set the micrometers at 450 µl, turn on the lights at the lowest intensity, and allow 10 to 20 minutes equilibration. Next, close the operating valves and, after 10 minutes, record the µl oxygen produced in each manometer. Without interruption, increase the light intensity at 10-minute intervals, and record the net µl oxygen produced per 10 minutes per flask at each intensity. At the end of the photosynthesis measurements repeat the measurement of dark respiration. Adjust the temperature of the water bath to 30°C and repeat the entire experiment. (NOTE: In practice, each laboratory section will perform the experiment at only one temperature; the data from all sections will be compiled and made available to all students.)

A variable transformer connected in the electrical circuit for the lights will be used to regulate light intensity at predetermined intensities. The variable transformer has been previously calibrated so that the settings indicated below provide approximately the designated light intensities at the bottoms of the flasks during operation.[3]

Setting on transformer (% maximum output voltage)	Approximate light intensity (ft-c)
46	50
53	100
62	200
71	400
84	800
100	1600

Plot two curves (one for 20°C and one for 30°C) on the same set of coordinates. Label the vertical axis (ordinate) as "Mean true rate of photosynthesis (µl O_2 produced/10 minutes);" label the horizontal axis (abscissa) as "Light intensity (ft-c)" and calibrate the abscissa from

---

[3]Each instrument should be calibrated when the variable transformer is first connected in the electrical circuit and occasionally thereafter as the lamps age and burned out lamps are replaced.

0 to 1600 ft-c. Plot both curves from 0 ft-c and label each curve to denote the temperature at which the data were obtained.

References

1. Arnon, D. I. 1967. Photosynthetic activity of isolated chloroplasts. Physiol. Rev. 47: 317-358.

2. Bishop, N. I. 1966. Partial reactions of photosynthesis and photoreduction. Ann. Rev. Plant Physiol. 17: 185-208.

3. Boardman, N. K. 1968. The photochemical system of photosynthesis. Adv. Enzymol. 30: 1-80.

4. Bonner, J. and A. W. Galston. 1952. Principles of Plant Physiology. W. H. Freeman and Company, San Francisco.

5. Calvin, M. 1962. The path of carbon in photosynthesis. Science 135: 879-889.

6. Calvin, M. and J. A. Bassham. 1962. The Photosynthesis of Carbon Compounds. W. A. Benjamin, New York.

7. Clayton, R. K. 1963. Photosynthesis: primary physical and chemical processes. Ann. Rev. Plant Physiol. 14: 159-180.

8. Clayton, R. K. 1965. The biophysical problems of photosynthesis. Science 149: 1346-1354.

9. Gibbs, M. 1967. Photosynthesis. Ann. Rev. Biochem. 36: 658-684.

10. Levine, R. P. 1969. The mechanism of photosynthesis. Scient. Am. 221 (No. 6, Dec.): 58-70.

11. Meyer, B. S., D. B. Anderson, and R. H. Böhning. 1960. Introduction to Plant Physiology. D. Van Nostrand Company, Princeton, New Jersey.

12. Rabinowitch, E. and Govindjee. 1969. Photosynthesis. John Wiley & Sons, New York.

13. Ray, P. M. 1972. The Living Plant. 2nd Ed. Holt, Rinehart and Winston, New York.

14. Salisbury, F. B. and C. W. Ross. 1978. Plant Physiology. 2nd Ed. Wadsworth Publishing Company, Belmont, California.

15. Talling, J. F. 1961. Photosynthesis under natural conditions. Ann. Rev. Plant Physiol. 12: 133-154.

16. Umbreit, W. W., R. H. Burris, and J. F. Stauffer. 1972. Manometric and Biochemical Techniques. 5th Ed. Burgess Publishing Company, Minneapolis.

17. Zelitch, I. 1971. Photosynthesis, Photorespiration, and Plant Productivity. Academic Press, New York.

Special Materials and Equipment Required
Per Laboratory Section

(1) Gilson Differential Respirometer (e.g. Gilson Model GRP-14, refrigerated, photosynthesis, 14-manometer model) or similar differential manometry apparatus, adapted for variable control of light intensity
(1) Light meter
*(Approximately 250 ml) Dense, bacteria-free liquid suspension of Scenedesmus obliquus cultured as described in text or purchased from a biological supply company (e.g. Carolina Biological Supply Company, Burlington, North Carolina 27215)
(Approximately 200 ml) Carbonate-bicarbonate buffer (0.1 M, pH 9)
(1) Clinical centrifuge

Recommendations for Scheduling

Half the prescribed experiment (light-saturation curve at one temperature) should be conducted as a class project in each laboratory section.

*Various other species of unicellular and colonial algae commonly available from biological supply companies, such as Chlorella, can be substituted for Scenedesmus obliquus. An inoculum of S. obliquus can be purchased from Culture Collection of Algae, Department of Botany, Indiana University, Bloomington, Indiana 47401, or from the American Type Culture Collection, 12301 Parklawn Drive, Rockville, Maryland 20852.

Duration: 1 laboratory period; exclusive attention to this experiment required throughout the period.

EXERCISE 5

EFFECTS OF TEMPERATURE AND LIGHT INTENSITY ON THE RATE OF PHOTOSYNTHESIS IN A GREEN ALGA

REPORT

Name _____ Section _____ Date _____

Dark respiration data:

Operating temperature: _____ °C

Time interval: _____ minutes

Flask No.	Before photosynthesis measurements			After photosynthesis measurements		
	Micrometer reading (µl)			Micrometer reading (µl)		
	Start	End	Δ	Start	End	Δ
1						
2						
3						
4						
5						
6						
7						
8						
9						
10						
11						
12						
13						
14						
Mean						

EFFECTS OF TEMPERATURE AND LIGHT INTENSITY ON THE RATE OF PHOTOSYNTHESIS IN A GREEN ALGA

REPORT

Name _____ Section _____ Date _____

Photosynthesis data:

Operating temperature: _____ °C

Time interval at each light intensity: _____ minutes

Flask no.	Zero time	50 ft-c	Δ	100	Δ	200	Δ	400	Δ	800	Δ	1600	Δ
Micrometer reading (µl)													
1													
2													
3													
4													
5													
6													
7													
8													
9													
10													
11													
12													
13													
14													
Mean Net Photosynthesis (µl/10 minutes)													
Mean Dark Respiration (µl/10 minutes)													
Mean True Photosynthesis (µl/10 minutes)													

EXERCISE 5

EFFECTS OF TEMPERATURE AND LIGHT INTENSITY ON THE RATE OF PHOTOSYNTHESIS IN A GREEN ALGA

REPORT

Name _____ Section _____ Date _____

Photosynthesis data:

Operating temperature: _____°C

Time interval at each light intensity: _____ minutes

Flask no.	Zero time	50 ft-c	Δ	100	Δ	200	Δ	400	Δ	800	Δ	1600	Δ

Micrometer reading (μl)

Flask no.	Zero time	50 ft-c	Δ	100	Δ	200	Δ	400	Δ	800	Δ	1600	Δ
1													
2													
3													
4													
5													
6													
7													
8													
9													
10													
11													
12													
13													
14													
Mean Net Photosynthesis (μl/10 minutes)													
Mean Dark Respiration (μl/10 minutes)													
Mean True Photosynthesis (μl/10 minutes)													

EFFECTS OF TEMPERATURE AND LIGHT INTENSITY ON THE RATE OF PHOTOSYNTHESIS IN A GREEN ALGA

REPORT

Name _____ Section _____ Date _____

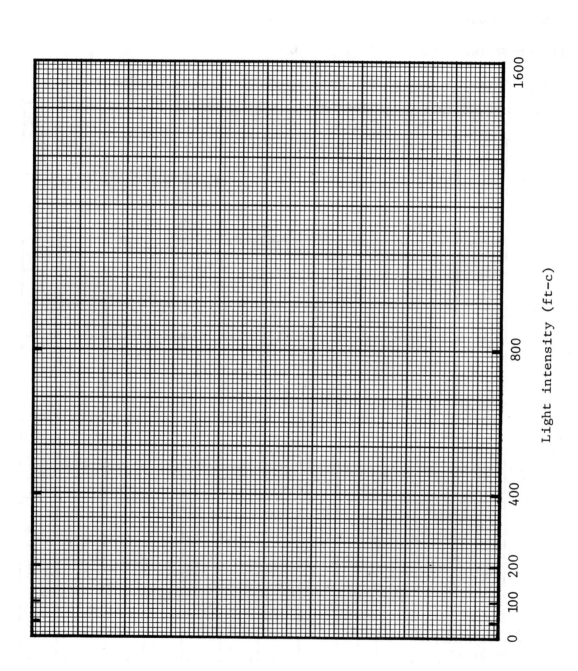

Light intensity (ft-c)

Mean true rate of photosynthesis
(μl O_2 produced/10 minutes)

Exercise 5

Effects of Temperature and Light Intensity on the Rate of Photosynthesis in a Green Alga

REPORT

Name _____ Section _____ Date _____

Questions

1. What was the temperature coefficient (20° to 30° C) for photosynthesis at 50 ft-c? at 1600 ft-c?

2. What do the results of this experiment indicate about the multiple-reaction nature of photosynthesis?

3. State the "principle of limiting factors" in a generally acceptable form, and explain how the principle was exemplified by this experiment.

EXERCISE 5

EFFECTS OF TEMPERATURE AND LIGHT INTENSITY ON THE RATE OF PHOTOSYNTHESIS IN A GREEN ALGA

REPORT

Name _____ Section _____ Date _____

4. List the three physical environmental factors which are of primary importance in the regulation of photosynthetic rate, and describe the circumstances under which each factor becomes rate-limiting for photosynthesis.

5. How is photosynthetic rate measured in terrestrial seed plants?

Exercise 6

Kinetics of Photosynthetic Carbon Dioxide Assimilation and Oxygen Evolution in a Green Alga

Introduction

Brief inspection of the empirical summary equation of photosynthesis reveals that one molecule of oxygen is produced for each molecule of carbon dioxide fixed:

$$CO_2 + 2H_2O \longrightarrow (CH_2O) + H_2O + O_2$$

The purpose of this experiment is to compare the kinetics of carbon dioxide assimilation and oxygen evolution in the alga Scenedesmus obliquus over a period of steady-state (constant rate) photosynthesis.

Materials and Methods[1]

The general procedure for setting up this experiment is identical to that described for Experiment 5 - "Effects of Temperature and Light Intensity on the Rate of Photosynthesis in a Green Alga" in this series. Refer to the "Materials and Methods" section of that exercise for information regarding procedures for culturing Scenedesmus.

Centrifuge down approximately 250 ml of the Scenedesmus culture provided. Resuspend the cells in 50 ml of carbonate-bicarbonate buffer (pH 9.08).[2] Pipet 3 ml of the dense algal suspension into the main compartment of each of the 14 reaction flasks for a Gilson respirometer. To each sidearm of each flask add 0.5 ml of carbonate-bicarbonate buffer containing 2.5 µc of $NaH^{14}CO_3$. Caution: Do not pipet the radioactive bicarbonate solution by mouth; use a propipet or other pipeting device and check with the instructor for specific procedures. Connect all the flasks to the Gilson respirometer. Keep the temperature of the water bath at a constant 30° C, and adjust the light intensity to the maximum attainable (approximately 1600 ft-c at the bottoms of the submerged flasks). Allow a period of 15 to 30 minutes for thermal equilibration of the flask contents with the water bath.

After equilibration, measure the rate of oxygen evolution in all the flasks at 10-minute intervals for a period of at least 30 minutes, or until a constant rate of oxygen evolution is attained. Open all the operating

---

[1]The assistance of Dr. Norman I. Bishop, Department of Botany and Plant Pathology, Oregon State University, Corvallis, with the design of basic procedures is gratefully acknowledged.

[2]Carbonate-bicarbonate buffer (pH 9.08): Mix 3 volumes of 0.1 M Na_2CO_3 and 17 volumes of 0.1 M $NaHCO_3$.

valves, adjust the micrometers to 450 µl, and again close the operating valves. Then tip the buffer containing $NaH^{14}CO_3$ from the sidearms of each flask. <u>Immediately</u> remove two of the flasks and harvest the <u>Scenedesmus</u> as described below. Thereafter, remove two flasks each at 5, 10, 15, 30, 45 and 60 minutes after tipping in the radioactive bicarbonate. <u>Record the readings on the micrometers connected to each pair of flasks at the time they are removed</u>.

To harvest the cells, pour the contents of the flask rapidly into a funnel fitted with a Millipore filter[3] and connected to a filtration flask and aspirator. Filter the cells rapidly under reduced pressure, and wash the cells three times with non-radioactive carbonate-bicarbonate buffer to rinse the $NaH^{14}CO_3$ off the surfaces of the cells. Quickly transfer the filter pad containing the cells to a liquid scintillation vial containing 10 to 15 ml of Bray's counting solution[4] or a commercially available counting solution for aqueous samples (e.g. Aquasol, New England Nuclear). Measure the radioactivity in each sample using a liquid scintillation counter. If time and circumstances permit, determine the counting efficiency for each sample and compute the actual disintegrations per minute (dpm). (Alternatively, radioactivity can be measured with a planchet counting system.)

On a common set of coordinates plot both the cumulative evolution of oxygen and the cumulative assimilation of ^{14}C versus time.

References

1. Arnon, D. I. 1960. The role of light in photosynthesis. Scient. Am. 203 (No. 5, Nov.): 104-118.

2. Bassham, J. A. 1962. The path of carbon in photosynthesis. Scient. Am. 206 (No. 6, June): 88-100.

3. Bray, G. A. 1960. A simple efficient liquid scintillator for counting aqueous solutions in a liquid scintillation counter. Anal. Biochem. 1: 279-285.

4. Calvin, M. 1962. The path of carbon in photosynthesis. Science 135: 879-889.

---

[3]The author uses a Millipore Microanalysis Filter Holder equipped with a Millipore HAWP 025 00 filter.

[4]The composition of Bray's solution:

Naphthalene	60.0 g
PPO (2,5-diphenyloxazole)	4.0 g
POPOP [1,4-<u>bis</u>-2'-(5'-phenyloxazolyl)-benzene]	0.2 g
Methanol (absolute)	100 ml
Ethylene glycol	20 ml
<u>para</u>-Dioxane to make	1 liter

5. Calvin, M. and J. A. Bassham. 1962. The Photosynthesis of Carbon Compounds. W. A. Benjamin, New York.

6. Lehninger, A. L. 1975. Biochemistry. 2nd Ed. Worth Publishers, New York.

7. Levine, R. P. 1969. The mechanism of photosynthesis. Scient. Am. 221 (No. 6, Dec.): 58-70.

8. Mahler, H. R. and E. H. Cordes. 1971. Biological Chemistry. 2nd Ed. Harper and Row Publishers, New York.

9. Rabinowitch, E. and Govindjee. 1969. Photosynthesis. John Wiley & Sons, New York.

10. Salisbury, F. B. and C. W. Ross. 1978. Plant Physiology. 2nd Ed. Wadsworth Publishing Company, Belmont, California.

11. Wang, C. H. and D. L. Willis. 1965. Radiotracer Methodology in Biological Science. Prentice-Hall, Englewood Cliffs, New Jersey.

12. Zelitch, I. 1971. Photosynthesis, Photorespiration, and Plant Productivity. Academic Press, New York.

Special Materials and Equipment Required Per Laboratory Section

(1) Gilson Differential Respirometer (e.g. Gilson Model GRP-14)
(Approximately 250 ml) Dense bacteria-free liquid suspension of Scenedesmus obliquus or other unicellular or colonial alga
(50 ml) Carbonate-bicarbonate buffer (0.1 M, pH 9)
(50 to 70 microcuries) Sodium bicarbonate-^{14}C (NaH^{14}CO$_3$)
(1 to several) Pipet fillers adaptable to 5- to 10-ml pipets
(1) Microfiltration apparatus (e.g. Millipore Microanalysis Filter Holder and Millipore HAWP 025 00 filters, available from Millipore Corporation, Bedford, Massachusetts 01730)
*(14) Liquid scintillation counting vials
*(150 to 300 ml) Bray's counting solution or commercially available counting solution for aqueous samples (e.g. Aquasol, available from New England Nuclear, 575 Albany Street, Boston, Massachusetts 02118, or nearest branch office)
*(1) Liquid scintillation spectrometer

Recommendations for Scheduling

Conduct as a class project in each laboratory section.

*Alternatively, filter pads may be dried and crudely counted with a planchet counting instrument.

Duration: 1 laboratory period; 2 hours exclusive attention required.

EXERCISE 6

KINETICS OF PHOTOSYNTHETIC CARBON DIOXIDE ASSIMILATION AND OXYGEN EVOLUTION IN A GREEN ALGA

REPORT

Name _____ Section _____ Date _____

Oxygen evolution after tip-in of $NaH^{14}CO_3$:

Flask no.	Time (minutes)	Micrometer reading (µl)		Net change (µl)	Mean change (µl)
		Initial	Final		
1	0	450	_____	_____	
2	0	450	_____	_____	_____
3	5	450	_____	_____	
4	5	450	_____	_____	_____
5	10	450	_____	_____	
6	10	450	_____	_____	_____
7	15	450	_____	_____	
8	15	450	_____	_____	_____
9	30	450	_____	_____	
10	30	450	_____	_____	_____
11	45	450	_____	_____	
12	45	450	_____	_____	_____
13	60	450	_____	_____	
14	60	450	_____	_____	_____

EXERCISE 6

KINETICS OF PHOTOSYNTHETIC CARBON DIOXIDE ASSIMILATION AND OXYGEN EVOLUTION IN A GREEN ALGA

REPORT

Name _____ Section _____ Date _____

^{14}C assimilation:

Flask no.	Time (minutes)	Net radioactivity[1] (cpm)	Mean radioactivity (cpm)	Mean radioactivity (dpm)
1	0	_____		
2	0	_____	_____	_____
3	5	_____		
4	5	_____	_____	_____
5	10	_____		
6	10	_____	_____	_____
7	15	_____		
8	15	_____	_____	_____
9	30	_____		
10	30	_____	_____	_____
11	45	_____		
12	45	_____	_____	_____
13	60	_____		
14	60	_____	_____	_____

[1]Minus background cpm.

EXERCISE 6

KINETICS OF PHOTOSYNTHETIC CARBON DIOXIDE ASSIMILATION AND OXYGEN EVOLUTION IN A GREEN ALGA

REPORT

Name _____ Section _____ Date _____

Graphical presentation of combined data:

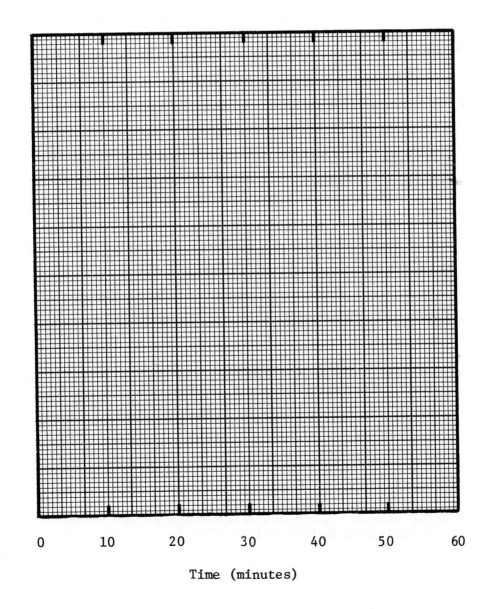

Exercise 6

Kinetics of Photosynthetic Carbon Dioxide Assimilation and Oxygen Evolution in a Green Alga

REPORT

Name _____ Section _____ Date _____

Questions

1. What do the results indicate specifically about the comparative kinetics of oxygen evolution and ^{14}carbon assimilation in Scenedesmus under the conditions of the experiment?

2. Describe the specific information that would be required to calculate the micromoles of oxygen evolved and micromoles of carbon dioxide (or HCO_3^-) assimilated during the experimental period. If possible, make these calculations, and compute the stoichiometry of oxygen evolution to carbon dioxide fixation.

3. Explain why the empirical summary equation for photosynthesis shows 2 molecules of water as reactant for each molecule of oxygen produced.

4. Describe briefly an experiment which would reveal whether the oxygen evolved in photosynthesis comes from carbon dioxide, water, or both reactants.

^{14}C-ASSAY OF PHOTORESPIRATION IN LEAF DISKS

Introduction

Quite possibly the most important single factor limiting net photosynthesis (defined as net carbon assimilated) are losses in dry weight due to respiratory metabolism. The familiar process of aerobic dark respiration consisting of the sequential reactions comprising glycolysis and the Krebs tricarboxylic acid cycle, and in many tissues also involving the oxidative pentose phosphate pathway, generally is well understood. This process of dark respiration, which, of course, occurs in all cells in both light and darkness, can itself cause a considerable decrease of dry weight gain. But in many species of chlorophyll-containing plants another, special kind of respiration--called photorespiration--occurs at a sufficiently high rate to alone account for a decrease of 50% of the gross photosynthetic assimilation (see Table 7-I).

Photorespiration may be defined as the light-dependent, aerobic respiration process in which glycolic acid, produced in photosynthesis, is oxidized to carbon dioxide in chlorophyll-containing tissues. This special kind of respiration thus is a distinctly different biochemical process from dark aerobic respiration. In addition to the fact that one type of respiration, but not the other, is dependent upon light, they differ in other important aspects: (a) as noted in the definition, glycolic acid, a product of photosynthetic carbon dioxide assimilation, is the substrate for photorespiration, while carbohydrates, lipids and proteins generally are considered the primary substances consumed in dark respiration; (b) coupled to dark respiration is conservation of energy in adenosine triphosphate (ATP), whereas energy released during glycolic acid oxidation is not conserved in a form useful to the cell; (c) the terminal sequence of reactions in aerobic dark respiration occurs in mitochondria, whereas glycolic acid, formed from an intermediate in photosynthetic assimilation of carbon dioxide in the chloroplast, is oxidized at least to glyoxylic acid in organelles called peroxisomes; (d) photorespiration is markedly stimulated by increasing concentrations of oxygen in the atmosphere, whereas dark respiration is fully operative at 2 to 3% O_2; (e) photorespiration increases more with increasing temperature than does dark respiration; and (f) photorespiration is inhibited by high CO_2 concentration (>1000 ppm), while dark respiration generally is not greatly affected by the CO_2 concentration.

The mechanism of photorespiration is incompletely elucidated at present, but several important features are known. The chloroplast is the site where glycolic acid is synthesized. This two-carbon substrate is thought to be derived from one or more sugar phosphates in the photosynthetic carbon reduction cycle through the action of a transketolase. However, the precise origin of glycolic acid is unknown, and there is some evidence that glycolic acid may be synthesized in more than one way. It is known also that the oxidation of glycolic acid to glyoxylic acid occurs in peroxisomes according to the following reaction:

$$CH_2OH\text{-}COOH \quad + \quad O_2 \quad \xrightarrow{\text{Glycolate oxidase}} \quad CHO\text{-}COOH \quad + \quad H_2O_2$$

Glycolic acid   Glyoxylic acid

In the absence of catalase, the glyoxylic acid can be oxidized nonenzymatically by the hydrogen peroxide to yield formic acid and CO_2, the CO_2 arising from the carboxyl-carbon of glycolic acid, as follows:

$$CHO\text{-}COOH \quad + \quad H_2O_2 \quad \longrightarrow \quad HCOOH \quad + \quad CO_2 \quad + \quad H_2O$$

Glyoxylic acid   Formic acid

The identity of the physiologically significant reaction by which glyoxylic acid is oxidized to CO_2, if different from the above, is not known. Purified peroxisomes do not oxidize glycolic beyond glyoxylic acid because of the large amount of catalase present in these particles. Whether glyoxylic acid oxidation to CO_2 occurs elsewhere in the cell or whether perhaps there is an enzymatic oxidation of glyoxylic acid to CO_2 is not known. Some investigators believe that glyoxylic acid must first be converted to glycine, and that CO_2 is released during the conversion of glycine to serine.

Generally, species may be classified as having either high or low rates of photorespiration. However, there are some species intermediate between the two extremes, and individual varieties of a single species (e.g. tobacco and corn) can show marked differences in rate of photorespiration. Examples of species with generally high rates of photorespiration are tobacco (Nicotiana tabacum), bean (Phaseolus vulgaris), wheat (Triticum aestivum), soybean (Glycine max), and sugar-beet (Beta vulgaris). Corn or maize (Zea mays), sugar-cane (Saccharum officinarum), sorghum (Sorghum vulgare), and some species of Amaranthus and Atriplex are examples of species with comparatively low rates of photorespiration. Several general differences between leaves of angiosperm species with high and low rates of photorespiration are given in Table 7-I. While this exercise concerns photorespiration in certain angiosperms, it should be noted that the process is also reported to occur in certain algae, liverworts, gymnosperms and other types of chlorophyllous plants.

Species which have low rates of photorespiration in general also exhibit the Hatch and Slack pathway of photosynthetic assimilation of carbon dioxide, and are commonly called "C_4-plants" to denote the formation of 4-carbon, dicarboxylic acids (aspartic and malic acids) as early products of photosynthesis. These plants are more efficient in terms of CO_2 assimilation per unit of light energy consumed than plants having high rates of photorespiration. The plants with low rates of photorespiration also generally produce more dry weight per unit of water transpired than those having high rates of photorespiration.

Plants which have high rates of photorespiration utilize only the Calvin-Benson cycle of photosynthetic assimilation of CO_2, and are commonly called "C_3-plants" to denote the formation of the 3-carbon phosphoglyceric acid as an early product of photosynthesis. These plants exhibit the Warburg effect, inhibition of photosynthesis by oxygen. In these plants, increasing oxygen concentrations up to an atmosphere of 100% O_2 increases the rate of photorespiration markedly, while plants with low rates of

Table 7-I. Some General Differences Between Leaves of
Species With High and Low Rates of Photorespiration[1]

Parameter or feature	High photorespiration	Low photorespiration
1. Rate of photorespiration, measured as CO_2 released	3 to 5 times dark respiration	<10% of dark respiration
2. Net rate of photosynthesis at 25°, bright light, 300 ppm CO_2 and 21% O_2 (mg dm^{-2} hr^{-1})	10 to 30	50 to 70
3. Light intensity for maximal rate of photosynthesis	1000 to 3000 ft-c	>10,000 ft-c
4. Effect of temperature (25° vs. 35°) on net photosynthesis	None, or lower rate at higher temperature	50 to 100% greater at higher temperature
5. Process of primary photosynthetic assimilation of CO_2	Calvin-Benson Cycle	Hatch-Slack Pathway
6. Anatomy and chloroplasts	No bundle sheath with specialized chloroplasts	Larger chloroplasts in the bundle sheaths
7. Rate of translocation of photosynthate	Slow	Rapid

[1]From Table 6.6, p. 211, in Zelitch (1971) with minor modifications.

photorespiration show no such response to oxygen. Whether it is this characteristic of photorespiration which accounts for the inhibition of net photosynthesis by O_2 in species showing the Warburg effect, or whether photorespiration is a result of the Warburg effect, is not clear.

Is photorespiration of value to plants? Conceivably, according to one view, glycolic acid formation might be a mechanism for removing a harmful substance generated by photooxidation from the chloroplasts. Under conditions favoring photorespiration and the Warburg effect--atmospheric concentration of O_2 and high light intensity--electrons generated in photochemical reactions of photosynthesis may be diverted away from the physiological acceptor nicotinamide adenine dinucleotide phosphate (NADP) resulting in the formation of H_2O_2. The H_2O_2 then could act as the agent oxidizing a two-carbon compound derived from the photosynthetic carbon reduction cycle to glycolic acid. Alternatively, electrons may be efficiently transferred to NADP, but the NADPH may be rapidly oxidized as glyoxylic acid (from peroxisomes) is transferred to chloroplasts and is reduced back to glycolic acid. However, there is no conclusive evidence that photorespiration really is of significant positive value. On the contrary, it is clear that net photosynthesis is reduced from gross photosynthesis to the extent that glycolic acid is oxidized to CO_2. Furthermore, in this oxidation the energy which is released is not conserved in a form useful to the cell. Hence, photorespiration may best be viewed as having negative consequences for the plant by virtue of reducing photosynthetic efficiency.

Photorespiration often occurs at rates three to five times greater than dark respiration. Since the former process is restricted to chlorophyll-containing tissues and to periods of photosynthesis, its overall effect on a daily basis (light plus darkness) may not be any greater than that of dark respiration which, of course, occurs in green and non-green tissues in both light and darkness. Nevertheless, photorespiration clearly is an important factor in limiting net photosynthesis in numerous species. It is experimentally demonstrable that significant increases in growth rates should be anticipated in plants with high rates of photorespiration if the process were somehow inhibited. Geneticists currently are undertaking breeding programs to develop varieties of crop plants which do not exhibit high rates of photorespiration.

The purpose of this experiment is to assay photorespiration in excised disks of tobacco (Nicotiana tabacum) and corn (Zea mays) using a sensitive ^{14}C-assay method.

Materials and Methods[2]

A. Fixation of $^{14}CO_2$. Cut off at the base several leaves from the tobacco and corn plants provided, and submerge the leaves basally in a container of water. Then cut 12 disks from symmetrical positions of the leaves of each species using a sharp cork borer or punch approximately 1.6 cm in diameter, and float the disks on water in petri dishes. Transfer with a forceps 6 disks of each species to two 100-ml Warburg reaction

---

[2]Adapted from: Zelitch, I. 1968. Investigations on photorespiration with a sensitive ^{14}C-assay. Plant Physiol. 43: 1829-1837.

vessels[3] each containing sufficient water to just cover the bottom. To the sidearm of each reaction vessel add 1.0 ml of aqueous solution containing 6 μmoles of nonradioactive $NaHCO_3$ and 2 μc of $NaH^{14}CO_3$ of known specific radioactivity. <u>Caution</u>: do not pipet the radioactive bicarbonate solution by mouth; use a propipet or other device and check with the instructor for specific procedure. Cap the venting plug with a rubber serum cap. Attach each reaction vessel to a Warburg manometer, leaving the stopcock open to the atmosphere, and submerse the reaction vessels in an illuminated constant-temperature water bath set at 30° or 35° C and providing a light intensity of approximately 1000 to 2000 ft-c incident uniformly on the flasks.[4] Allow 45 minutes to insure that the stomates of the leaf disks become fully open and thermal equilibration between the reaction vessels and the water bath is achieved. Then stop the shaker, close the stopcock on each manometer, and seal the other, open arm of the manometer with a serum cap or rubber stopper. With a syringe inject 0.1 ml of 0.1 N HCl (= 10 μmoles HCl) through the serum cap into the sidearm of each to liberate CO_2. Again turn on the shaker and allow assimilation of CO_2 to proceed for approximately 30 minutes before starting the measurements of photorespiration.

   B. <u>Photorespiration assay</u>. Connect a length of rubber or plastic tubing to an air jet, pressure pump, or compressed cylinder of air. Then, using tubing, rubber stoppers, glass tubing connections, 500-ml Erlenmeyer flasks, rig apparatus with which to pass the air first through 250 ml of 2.5 N KOH and then through 250 ml of water. Using glass or plastic Ys or Ts and tubing, divide the effluent stream of CO_2-free air and prepare a length of tubing for connection to the top of the arm bearing the stopcock on each manometer. Adjust the flow rate of the CO_2-free air to go

---

[3] Other apparatus could readily be substituted for that which is described; the experiment could be scaled down and performed satisfactorily with conventional photosynthesis models of differential respirometers (e.g. a Gilson Respirometer - refrigerated, photosynthesis model; see Exercise 5). To use the Gilson Respirometer with the small reaction vessels (approximately 14 to 17 ml capacity), the following modifications in procedure should be noted: (1) disconnect the Tygon tubing leading from each reaction vessel to the manometer; (2) cut a single "doughnut-shaped" sample of tissue to fit over the center well (if present) and nearly cover the bottom of each reaction vessel; (3) put approximately 0.5 to 1.0 ml (containing about 2 μmoles of nonradioactive $NaHCO_3$ and 1 μc of $NaH^{14}CO_3$) in one sidearm of each reaction vessel (reduce amount of HCl proportionately); (4) use the right-angle glass connector which inserts into the top of the reaction vessel as the influent for CO_2-free air, and use the venting plug of the empty sidearm as the effluent; (5) attach reaction vessels to shaking apparatus, and submerse in constant temperature bath during operation.

[4] If available, the illuminated water bath of a photosynthesis model of Warburg apparatus or Gilson respirometer can be used. Otherwise, a common thermoregulated water bath, tungsten lamps, and mechanical shaker can be rigged to work satisfactorily. For best results the reaction vessels should be shaken at a moderate rate throughout the incubation.

to each manometer to approximately 500 ml per minute. Replace the serum cap on the sidearm of each reaction vessel with a venting plug. Prepare for attachment to the venting plug on each reaction vessel another length of tubing the distal end of which can be fitted with an aerator and submersed in a flask containing 50 ml of 1.0 M ethanolamine (to trap the $^{14}CO_2$ released). Prepare the ethanolamine traps, and also prepare 28 liquid scintillation vials each containing 10 to 15 ml of Bray's counting solution.

Connect all tubing, submerge effluent tubing from all four reaction vessels in a common flask containing about 50 ml of 2.5 N KOH (this to be discarded later), open the stopcock on the manometers, remove the serum cap from the other, open end of the manometer, remove the serum caps from the venting plugs, turn on the stream of CO_2-free air and flush the reaction vessels of any unassimilated $^{14}CO_2$ for approximately five minutes. Then quickly submerge the effluent tubing from each reaction vessel in an ethanolamine trap and immediately ("zero time") withdraw a 0.25 ml sample from each ethanolamine trap (do not pipet by mouth) and place the sample in a liquid scintillation vial. Leave the CO_2-free stream of air moving through the reaction vessels and into the ethanolamine traps, and take a 0.25-ml sample from each also at 10, 20, and 30 minutes.

C. Measurements of dark respiration. After assaying photorespiration for 30 minutes, turn off the lights illuminating the reaction vessels, and drape black cloth over the constant temperature bath. Trap the effluent $^{14}CO_2$ from all flasks in the common flask containing 50 ml of 2.5 N KOH (which was used previously, and which is to be discarded at the end of the experiment) during the first 15 minutes in darkness. Then resume the passing of CO_2-free air through the reaction vessels and sampling the original ethanolamine $^{14}CO_2$-trap solutions at 10-minute intervals for a total duration of 30 minutes.

Measure the radioactivity in all the samples of trapped $^{14}CO_2$ by liquid scintillation counting. Record the radioactivity data (in counts per minute, cpm) in a table. Plot a separate graph for each species, using the average values for the duplicate samples. Calculate for each species the ratio of $^{14}CO_2$ release in light to $^{14}CO_2$ release in darkness by dividing the average total cpm in samples collected in 30 minutes in the light by the average total cpm in samples collected in 30 minutes in darkness.

References

1. Bray, G. A. 1960. A simple efficient liquid scintillator for counting aqueous solutions in a liquid scintillation counter. Anal. Biochem. 1: 279-285.

2. Bull, T. A. 1969. Photosynthetic efficiencies and photorespiration in Calvin cycle and C_4-dicarboxylic acid plants. Crop Sci. 9: 726-729.

3. Decker, J. P. 1970. Early History of Photorespiration. Bioeng. Bull. No. 10. Eng. Res. Center, Arizona State University, Tempe, Arizona.

4. Ellyard, P. W. and M. Gibbs. 1969. Inhibition of photosynthesis by oxygen in isolated spinach chloroplasts. Plant Physiol. 44: 1115-1121.

5. Forrester, M. L., G. Krotkov, and C. D. Nelson. 1966a. Effect of oxygen on photosynthesis, photorespiration and respiration in detached leaves. I. Soybean. Plant Physiol. 41: 422-427.

6. Forrester, M. L., G. Krotkov, and C. D. Nelson. 1966b. Effect of oxygen on photosynthesis, photorespiration and respiration in detached leaves. II. Corn and other monocotyledons. Plant Physiol. 41: 428-431.

7. Gibbs, M. 1970a. Photorespiration, Warburg effect and glycolate. Ann. New York Acad. Sci. 168: 356-368.

8. Gibbs, M. 1970b. The inhibition of photosynthesis by oxygen. Am. Scientist 58: 634-640.

9. Goldsworthy, A. 1966. Experiments on the origin of CO_2 released by tobacco leaf segments in the light. Phytochem. 5: 1013-1019.

10. Goldsworthy, A. 1968. Comparison of the kinetics of photosynthetic carbon dioxide fixation in maize, sugar cane and tobacco, and its relation to photorespiration. Nature 217: 62.

11. Goldsworthy, A. 1970. Photorespiration. Botan. Rev. 36: 321-340.

12. Hatch, M. D., C. B. Osmond, and R. O. Slatyer. 1971. Photosynthesis and Photorespiration. John Wiley and Sons, New York.

13. Hough, R. A. and R. G. Wetzel. 1972. A ^{14}C-assay for photorespiration in aquatic plants. Plant Physiol. 49: 987-990.

14. Jackson, W. A. and R. J. Volk. 1970. Photorespiration. Ann. Rev. Plant Physiol. 21: 385-432.

15. Menz, K. M., D. N. Moss, R. Q. Cannell, and W. A. Brun. 1969. Screening for photosynthetic efficiency. Crop Sci. 9: 692-694.

16. Poskuta, G., C. D. Nelson, and G. Krotkov. 1967. Effects of metabolic inhibitors on the rates of CO_2 evolution in light and in darkness by detached spruce twigs, wheat and soybean leaves. Plant Physiol. 42: 1187-1190.

17. Salisbury, F. B. and C. W. Ross. 1978. Plant Physiology. 2nd Ed. Wadsworth Publishing Company, Belmont, California.

18. Tolbert, N. E. and R. K. Yamazaki. 1970. Leaf peroxisomes and their relation to photorespiration and photosynthesis. Ann. New York Acad. Sci. 168: 325-341.

19. Tregunna, E. B., G. Krotkov, and C. D. Nelson. 1966. Effect of oxygen on the rate of photorespiration in detached tobacco leaves. Physiol. Plantarum 19: 723-733.

20. Tregunna, E. B., B. N. Smith, J. A. Berry, and W. J. S. Downton. 1970. Some methods for studying the photosynthetic taxonomy of the angiosperms. Can. J. Botany 48: 1209-1214.

21. Wang, C. H. and D. L. Willis. 1965. Radiotracer Methodology in Biological Science. Prentice-Hall, Englewood Cliffs, New Jersey.

22. Zelitch, I. 1959. The relationship of glycolic acid to respiration and photosynthesis in tobacco leaves. J. Biol. Chem. 234: 3077-3081.

23. Zelitch, I. 1964. Organic acids and respiration in photosynthetic tissues. Ann. Rev. Plant Physiol. 15: 121-142.

24. Zelitch, I. 1966. Increased rate of net photosynthetic carbon dioxide uptake caused by the inhibition of glycolate oxidase. Plant Physiol. 41: 1623-1631.

25. Zelitch, I. 1968. Investigations on photorespiration with a sensitive ^{14}C-assay. Plant Physiol. 43: 1829-1837.

26. Zelitch, I. 1971. Photosynthesis, Photorespiration, and Plant Productivity. Academic Press, New York.

Special Materials and Equipment Required
Per Laboratory Section

(1 or 2) Vigorous tobacco (<u>Nicotiana tabacum</u>) plant, e.g. variety Havana Seed

(1 or 2) Vigorous corn (<u>Zea mays</u> plant, e.g. hybrid Penn 602A

*(4) Large (75- to 100-ml) Warburg reaction vessels each with a sidearm and venting plug (available in 100-ml size from AMINCO, 8030 Georgia Avenue, Silver Spring, Maryland 20910)

*(4) Warburg manometers

*(1) Illuminated, thermoregulated water bath

*(1) Shaking apparatus

(1) Source of air pressure

(20 μmoles) $NaHCO_3$

(4 to 8 μc) $NaH^{14}CO_3$

(200 ml) 1.0 M ethanolamine solution

---

*See footnote 3.

(28) Liquid scintillation vials
(280 to 420 ml) Bray's liquid scintillation counting solution (see
    Exercise 6 or reference 1) or commercially available counting
    solution for aqueous samples (e.g. Aquasol, available from New
    England Nuclear, 575 Albany Street, Boston, Massachusetts 02118)
(1 to several) Pipet fillers to fit 0.25- to 1.0-ml pipets
(1 or 2) 0.1-ml syringes
(1 to several) Cork borers, 1.6 cm diameter
(300 ml) 2.5 N KOH
(Approximately 6 to 10 feet) Rubber or Tygon tubing

Recommendations for Scheduling

This experiment will, in most cases, have to be conducted as a class
project in each laboratory section. The tobacco and corn plants, of
course, should be grown in advance and be available in the laboratory on
the day the experiment is to be performed. Other advisable advance preparations are to: collect all necessary apparatus; warm the constant
temperature water bath to 30° or 35°; and determine how to adjust the
flow of CO_2-free air to approximately 500 ml per minute. The liquid
scintillation counting can either be done by the instructor or by student
volunteers after the scheduled laboratory period and the data presented
to all the students at a subsequent class meeting.

Duration: One 2- to 3-hour laboratory period; exclusive attention
to this experiment required.

EXERCISE 7

^{14}C-ASSAY OF PHOTORESPIRATION IN LEAF DISKS

REPORT

Name _____ Section _____ Date _____

Measurements of $^{14}CO_2$ released:

| Time (minutes) | Radioactivity/0.25 sample (cpm) | | | | | |
	Tobacco Flask 1	Tobacco Flask 2	Mean	Corn Flask 1	Corn Flask 2	Mean
			Photorespiration			
0	_____	_____	_____	_____	_____	_____
10	_____	_____	_____	_____	_____	_____
20	_____	_____	_____	_____	_____	_____
30	_____	_____	_____	_____	_____	_____
Σ 10 to 30	_____	_____	_____	_____	_____	_____
			Dark Respiration			
10	_____	_____	_____	_____	_____	_____
20	_____	_____	_____	_____	_____	_____
30	_____	_____	_____	_____	_____	_____
Σ 10 to 30	_____	_____	_____	_____	_____	_____

EXERCISE 7

^{14}C-ASSAY OF PHOTORESPIRATION IN LEAF DISKS

REPORT

Name _____ Section _____ Date _____

Calculations of ratios of amounts of $^{14}CO_2$ released in the light and in darkness:

Tobacco: Corn:

Time courses of $^{14}CO_2$ release:

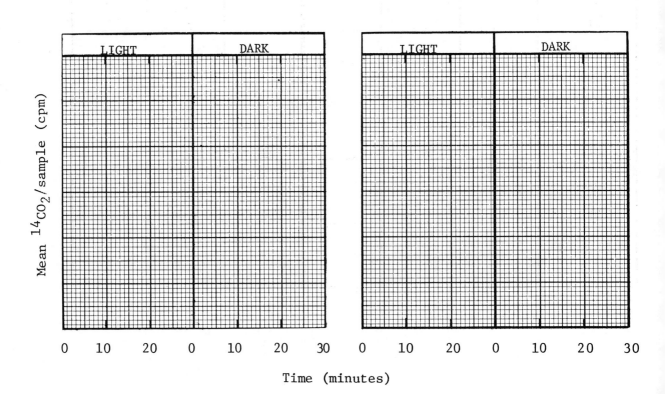

Tobacco   Corn

Mean $^{14}CO_2$/sample (cpm)

Time (minutes)

EXERCISE 7

^{14}C-ASSAY OF PHOTORESPIRATION IN LEAF DISKS

REPORT

Name _____ Section _____ Date _____

Questions

1. What conclusions can be made from this experiment concerning the comparative rates of photorespiration, if any, in the excised leaf disks of tobacco and corn?

2. Was it possible by the procedure employed in this experiment to measure gross photorespiration? Explain fully.

3. Why was CO_2-free air instead of normal air (containing 300 ppm CO_2) passed rapidly over the leaf disks during the assay of photorespiration?

EXERCISE 7

^{14}C-ASSAY OF PHOTORESPIRATION IN LEAF DISKS

REPORT

Name _____ Section _____ Date _____

4. What would be the predictable effect of passing a stream of 100% O_2 instead of (CO_2-free) air over the leaf disks during the photorespiration assay?

5. Describe the effect that α-hydroxy-2-pyridinemethanesulfonic acid, an inhibitor of glycolic acid oxidase, would be expected to have on the rate of photorespiration.

6. Briefly describe other methods, besides the ^{14}C-assay method used in this experiment, for assaying photorespiration.

EXERCISE 8

POLYACRYLAMIDE GEL ELECTROPHORESIS OF PLANT PROTEINS[1]

Introduction

Development is characterized by an ordered, controlled sequence of changes that occur within an organism at the molecular, cell, tissue and organ levels. Differentiation of cells and tissues involves processes that give rise to a variety of cell types within different tissues. Such cells within a particular tissue or organ are characterized by a unique morphological and chemical makeup.

Steward, Lyndon and Barber (1965) demonstrated that different parts of a pea seedling all contain different protein complements. Presumably, all cells within these different tissues have the same genetic constitution and therefore the same DNA encoded information in their chromosomes. Thus, if all cells contain the information necessary for the manufacture of all the different types of proteins found in the pea plant, what is it that controls the expression of this information? How does gene activity result in one protein complement in one part of the plant and an entirely different complement in another part? What controls the appearance of specific proteins in an organ at particular developmental stages? These are some major developmental questions that need answering.

The fact that different tissues within a plant contain unique complements of proteins was demonstrated by the above investigators using the elegant technique of polyacrylamide gel electrophoresis (Fig. 8-1). In this technique, a gel composed of polyacrylamide is polymerized inside a glass tube, across which is imposed an electrical potential which will cause any appropriately charged macromolecule present to migrate through the gel (Fig. 8-2). The preparation of these gels is a simple, rapid procedure in which the resulting pore size of the gel can easily be adjusted. The separation of nucleic acids and polysaccharides, as well as proteins, occurs on the basis of molecular size and charge.

The original theoretical and technical framework upon which this procedure rests, was developed about 17 years ago by Leonard Ornstein and Baruch J. Davis. It is based on well-known electrochemical laws as formulated in the Kohlrausch Regulating Function (see paper by Ornstein in reference 14 for complete details). If two ion solutions of the same charge but each possessing different mobilities in an electric field, are layered one on top of the other (the lower solution being the denser and faster of the two ions), the boundary between the two ionic species is sharply maintained as they migrate in an electric field. If a protein solution is sandwiched between the faster moving ion below (e.g. chloride), and a slower moving ion above (e.g. glycine), the proteins also migrate as a sharp band at the boundary of the two ions.

---

[1]Contributed by Dr. Ralph S. Quatrano, Department of Botany and Plant Pathology, Oregon State University, Corvallis, to whom grateful acknowledgment is expressed.

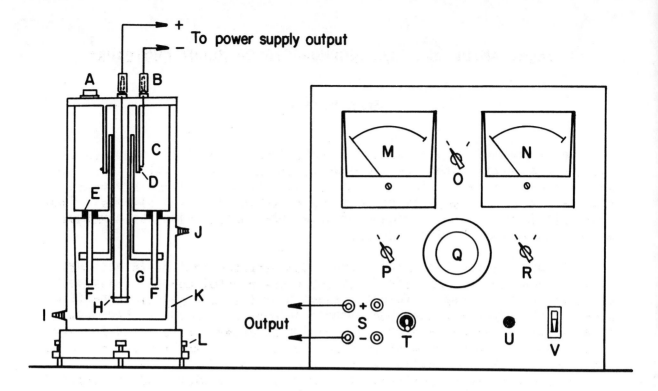

Fig. 8-1. Diagram of Buchler Analytical Polyacrylamide Vertical Disc Gel
Electrophoresis Apparatus and Buchler Constant D.C. Current and Constant
Voltage Power Supply.

A	Bubble level	L	Leveling screw
B	Electrode terminals	M	Voltage meter
C	Upper buffer chamber	N	Milliammeter
D	Upper electrode (platinum wire)	O	Voltage/current regulation selector
E	Sealing grommet		
F	Gel columns (12)	P	Voltage range selector (Hi/Lo)
G	Lower buffer chamber	Q	Voltage-current regulator knob
H	Lower electrode (platinum wire)	R	Current range selector (Hi/Lo)
I	Cooling water inlet	S	Output terminals (2 sets)
J	Cooling water outlet	T	Polarity selector switch
K	Water jacket	U	Pilot lamp
		V	Main power switch

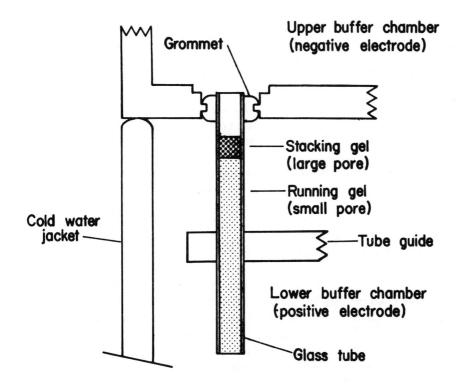

Fig. 8-2. Diagram of a single gel column mounted in the electrophoresis apparatus.

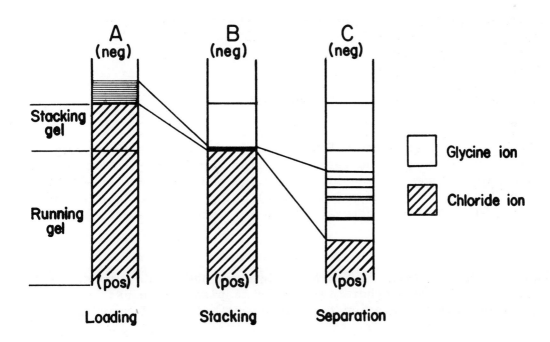

Fig. 8-3. Schematic diagrams of the behavior of proteins and other charged particles during electrophoresis as described in the text.

Most proteins have a net negative charge and free mobility toward the anode above pH 8.0. If the protein sample that you wish to separate is in a glycine buffer (pH 8.3) and is layered over a large-pore polyacrylamide gel (stacking gel), which is at the same pH but contains chloride ions, then a glycine-chloride boundary is formed (Fig. 8-3A). When a voltage gradient is applied, the proteins will move faster than the glycine ions but not as fast as the chloride ions. In a matter of time, the proteins will be moving as a sharp boundary between the two ionic species through the stacking gel. As these proteins migrate through the large-pore gel, each protein species is concentrated at the boundary in very thin discs ($\sim 25\mu$), one on top of the other, in order of decreasing mobility with the last followed by glycine (Fig. 8-3B). This steady-state stacking occurs within minutes after the voltage is applied, and one can actually observe this phenomenon by watching the concentration of some sample dye solution (e.g. bromphenol blue) into an extremely thin disc.

If the thin starting layer composed of many concentrated protein discs now passes into a gel with small pores (which would retard the protein molecules) and/or a higher pH (which would increase the mobility of glycine), the glycine ion will overtake the proteins and run directly behind the chloride ion. The trailing proteins would now be in a uniform, linear voltage gradient as well as in highly concentrated, thin starting zones. Each protein is now free to migrate through the running gel, and a mixture of proteins will separate from each other on the basis of their molecular sizes and charges (Fig. 8-3C). After a sufficient time to achieve satisfactory separation of the protein discs in the running gel, the proteins are immobilized and stained by placing the gels in an acid-dye solution. After leaching of the unbound stain, the components of a particular protein extract can be visualized in the gel.

It is the purpose of this experiment to demonstrate the presence of differences among the proteins in different organs of 3 to 5-day-old etiolated pea seedlings, and to acquaint you with the technique of poly-acrylamide gel electrophoresis.

Materials and Methods[2]

A. Preparation of extracts. Homogenize frozen samples (5 to 10 g) of hypocotyl-root, epicotyl, and cotyledons from 3 to 5-day-old etiolated Alaska pea (Pisum sativum L.) seedlings in 0.1 M phosphate buffer - 0.25 M sucrose, pH 7.2[3] (1 ml buffer : 1 g tissue) using a chilled mortar and pestle. Centrifuge the homogenate at 17,000 x g for 15 minutes, and

---

[2]Adapted from: Steward, F. C., R. F. Lyndon, and J. T. Barber. 1965. Acrylamide gel electrophoresis of soluble plant proteins: a study on pea seedlings in relation to development. Am. J. Botany 52: 155-164.

[3]Phosphate buffer, 0.1 M, pH 7.2: Combine 28 ml of 0.2 M NaH_2PO_4 and 72 ml of 0.2 M Na_2HPO_4, add and dissolve 17.1 g sucrose and dilute to a total volume of 200 ml with distilled water.

dialyze the supernatant overnight at 4° C against several changes of 0.1 M Tris-glycine buffer, pH 8.3.[4] Then centrifuge the dialysate at 39,000 x g for 30 minutes. Discard the pellet and add solid sucrose to the supernatant to a final concentration of 0.5 M. If possible, determine the protein content of each extract [e.g. by the method of Lowry et al. 1951. J. Biol. Chem. 193: 265–275] and calculate the volume of each extract necessary to give approximately 300 µg protein; if the protein concentration exceeds 300 µg/0.1 ml, dilute with 0.5 M sucrose solution to about that concentration. These solutions are the crude protein extracts to be electrophoresed.[5] Approximately 300 µg of protein in about 0.1 ml is a desirable-size sample to apply to each gel.

B. _Preparation of the gels_. Make up stock solutions and prepare gels as follows:

1. _Stock solutions_

   a. _Running gel (lower gel)_

      Solution A
1.0 N HCl	24.0 ml
Tris (hydroxymethyl) aminomethane (TRIS)	15.85 g
N,N,N',N'-Tetramethylethylenediamine (TEMED)	0.32 ml

      Make volume up to 100 ml with distilled water.

      Solution B
Acrylamide (Caution!)	30.0 g
N,N'-Methylenebisacrylamide	0.8 g

      Make volume up to 100 ml with distilled water.

      Solution C – Make fresh each time.
Ammonium persulfate	0.14 g/100 ml distilled water

   b. _Stacking gel (upper gel)_

      Solution D – Make fresh each time.
Ammonium persulfate	0.56 g/100 ml distilled water

------

[4]Tris-glycine buffer, 0.1 M, pH 8.3: Dissolve 12 g of Tris and 56.8 g of glycine in distilled water and dilute further to a total volume of 1 liter.

[5]If it is not possible to analyze the protein contents of the extracts in advance, satisfactory results still should be achieved providing extracts are prepared as indicated and different volumes of each extract are electrophoresed. Applications of 0.02 to 0.15 ml (e.g. 0.02, 0.05, 0.10 and 0.15 ml) of each dialyzed extract to different gels will insure appropriate concentrations of proteins for comparing banding patterns.

Solution E

1 N HCl	48.0 ml
TRIS	5.98 g
TEMED	0.46 ml

Make volume up to 100 ml with distilled water.

Solution F

Acrylamide (Caution!)	10.5 g
N,N'-Methylenebisacrylamide	2.5 g

Make volume up to 100 ml with distilled water.

Solution G

1.0 M sucrose (34.2 g diluted to 100 ml with distilled
water)

All stock solutions should be stored in brown bottles
in the cold (4° C)

2. Working solutions

a. Running gel (7.5% acrylamide, small pore size – makes
24 small tubes) – Make fresh each time, and bring stock
solutions to room temperature before mixing to avoid
formation of bubbles during polymerization.

1 part A		(10 ml)
1 part B	mix	(10 ml)
2 parts C	Mix gently by swirling.	(20 ml)

Fill glass tubes (7.5 cm long) stoppered at one end with serum caps
and supported in vertical position in a rack with the above running gel
solution to a height of 5.0-5.5 cm using a disposable pipet. Layer distilled water (3 mm) on top of the gels with a disposable pipet drawn out
to a fine point. Apply drops of water carefully so as not to mix the
water with the gel solution. Allow the gels to set 30-60 minutes. Then
shake the water from the tops of the tubes and remove the final drop of
water with a thin filter paper strip.

b. Stacking gel (2.6% acrylamide, large pore size – for
24 small tubes) – Make fresh each time, and bring to
room temperature before mixing to avoid formation of
bubbles during polymerization.

1 part D	(2 ml)
1 part E	(2 ml)
2 parts F	(4 ml)
4 parts G	(8 ml)

Apply the stacking gel on top of the polymerized running gel in each
tube. Again carefully layer water on top of the 0.5 cm stacking gel.
Then allow the gels to set for 15 minutes or until they appear cloudy.

C. Electrophoresis. When the gels have polymerized, shake the water from the top of the gels, and carefully remove the serum caps from the bottoms of the tubes. Place the tubes in the top part of the electrophoresis apparatus. Apply protein extract (0.02 to 0.15 ml; see part A) containing 0.5 M sucrose to each gel. Carefully layer Tris-glycine buffer (pH 8.3, 0.1 M) on top of the protein sample to fill the tube. A sharp interface between the gel and the protein and between the protein and the buffer is required. Layer a drop or two of bromophenol blue dye solution (dense aqueous solution) containing sucrose in the Tris-glycine buffer to one gel for a dye marker.

Add the Tris-glycine buffer to the lower part of the electrophoresis apparatus. This must be deep enough to cover the bottom terminal and to allow the gel tubes to extend into the lower buffer. Suspend a drop of buffer from the end of each tube to insure that no air bubbles are trapped. Then fit the upper part of the electrophoresis unit to the lower part and fill the upper reservoir with buffer to cover the gel tubes and to cover the top terminal.

Connect the positive (+) or red electrode to the bottom and the negative (−) or black electrode on the top. Start the electrophoresis run with a current of 2 ma/tube and increase to 4 ma/tube as soon as the dye marker is into the running gel. Conduct the electrophoresis at room temperature for about one hour, or until the dye marker is approximately 1 cm from the bottom of the gel.

D. Staining. At the end of the electrophoresis time, remove the gels from the tubes with a fine stream of water. Immediately immerse them in 0.7% Buffalo Black in 7.0% acetic acid and stain for about 1 to 5 hours at room temperature. Then destain the gels by placing them in 7% acetic acid. Change the acetic acid solution several times until the solution is no longer blue.

Present the results of this exercise as a series of detailed diagrams of the destained gels drawn to scale. If a spectrodensitometer is available, also scan the gels and present copies of the densitometric tracings in your report.

References

1. Bonner, J. 1965. The Molecular Biology of Development. Oxford University Press, Oxford and New York.

2. Bonner, J., M. E. Dahmus, D. Fambrough, R. C. Huang, K. Marushige, and D. Y. H. Tuan. 1968. The biology of isolated chromatin. Science 159: 47-56.

3. Chrambach, A. and D. Rodbard. 1971. Polyacrylamide gel electrophoresis. Science 172: 440-451.

4. Galston, A. W. and P. J. Davies. 1970. Control Mechanisms in Plant Development. Prentice-Hall, Englewood Cliffs, New Jersey.

5. Gordon, A. H. 1969. Electrophoresis of proteins in polyacryl-amide and starch gels. Pp. 1-149 in: T. S. Work and E. Work, Eds. Laboratory Techniques in Biochemistry and Molecular Biology. Vol. 1. Elsevier Publishing Company, New York.

6. Gross, P. R. 1968. Biochemistry of differentiation. Ann. Rev. Biochem. 37: 631-660.

7. Moore, T. C. 1979. Biochemistry and Physiology of Plant Hormones. Springer-Verlag, New York.

8. Quatrano, R. S. 1974. Development in marine organisms. Pp. 303-346 in: R. Mariscal, Ed. Experimental Marine Biology. Academic Press, New York.

9. Salisbury, F. B. and C. W. Ross. 1978. Plant Physiology. 2nd Ed. Wadsworth Publishing Company, Belmont, California.

10. Shaw, D. J. 1969. Electrophoresis. Academic Press, New York.

11. Steward, F. C. 1968. Growth and Organization in Plants. Addison-Wesley Publishing Company, Reading, Massachusetts.

12. Steward, F. C., R. F. Lyndon, and J. T. Barber. 1965. Acrylamide gel electrophoresis of souble plant proteins: a study on pea seedlings in relation to development. Am. J. Botany 52: 155-164.

13. Torrey, J. G. 1967. Development in Flowering Plants. Macmillan Company, New York.

14. Wareing, P. F. and I. D. J. Phillips. 1970. The Control of Growth and Differentiation in Plants. Pergamon Press, New York.

15. Whipple, H. E., Ed. 1964. Gel electrophoresis. Ann. New York Acad. Sci. 121: 305-650 (Article 2).

Special Materials and Equipment Required
Per Laboratory Section

(1) Analytical vertical disc gel electrophoresis apparatus with power supply [e.g. Buchler Polyanalyst and Buchler Power Supply (regulated constant D. C. current and constant voltage), available from Buchler Instruments, Incorporated, 1327 16th Street, Fort Lee, New Jersey 07024]
(Approximately 100) Etiolated Alaska pea seedlings, 3 days old
(Approximately 250 ml) Phosphate buffer (0.1 M, pH 7.2) containing 0.5 M sucrose
(Approximately 2 liters) Tris-glycine buffer (0.1 M, pH 8.3)
(100 ml each) Stock solutions A-G, described in Materials and Methods

(12) Glass electrophoresis tubes (5 mm I.D., 8 mm O.D., 7.5 cm long) and
     12 serum caps
(Approximately 10 ml) Bromophenol blue dye solution
(Approximately 2 liters) Buffalo Black dye solution
(Approximately 2 liters) 7% acetic acid

Recommendations for Scheduling

Conduct as a demonstration in each laboratory section. Pea seedlings should be grown and extracts prepared and dialyzed in advance of
the start of the laboratory period when the demonstration is to be performed. If possible, the protein concentrations of the extracts should
also be determined in advance and diluted if necessary (see footnote 5).
Buffers and dye solutions also should be prepared in advance.

Duration: 2 laboratory periods; all of first period, 30 minutes to
1 hour second period.

EXERCISE 8

POLYACRYLAMIDE GEL ELECTROPHORESIS OF PLANT PROTEINS

REPORT

Name _____ Section _____ Date _____

Diagrams of destained gels:

	Top (cathode) (−)	Bottom (anode) (+)
Identification of gel		

POLYACRYLAMIDE GEL ELECTROPHORESIS OF PLANT PROTEINS

REPORT

Name _____ Section _____ Date _____

Questions

1. Describe the similarities and differences among the electrophoretic banding patterns obtained with extracts of the different parts of the pea seedlings. Were there bands apparently common to all the gels? Did similarities exceed differences among the banding patterns? Explain.

2. Is it probable that each visible band of stained protein on each gel contained only one molecular species? Discuss.

3. Is it likely that all the soluble proteins present in the extract migrated in the gels, under the conditions which were employed? Explain.

POLYACRYLAMIDE GEL ELECTROPHORESIS OF PLANT PROTEINS

REPORT

Name _____ Section_____ Date _____

4. Are the observed differences among the protein complements from the different organs of the pea seedlings primarily the <u>cause</u> or the <u>result</u> of differentiation? Discuss.

5. What are some other possible applications of polyacrylamide gel electrophoresis besides analysis cf soluble proteins?

6. How can the results of analytical polyacrylamide electrophoresis be made quantitative?

7. Explain the difference between analytical and preparative polyacrylamide electrophoresis. What advantage does the latter procedure afford?

EXERCISE 9

EFFECTS OF AUXIN AND CYTOKININ ON MORPHOGENESIS
IN CALLUS TISSUE

Introduction

Development, as applied to plants arising by sexual reproduction, refers to the sum of the gradual and progressive qualitative and quantitative changes which comprise the transformation of a zygote into a mature, reproductive plant. The phenomenon is characterized by changes in size and weight, appearances of new structures and functions and losses of former ones, and temporary spatial and temporal discontinuities and changes in rate. Of course it is quite appropriate to speak of development of individual organs as well as whole plants.

A useful concept is to view development as consisting of three processes which normally occur concomitantly - growth, cellular differentiation and morphogenesis. Growth refers to the quantitative changes and may be defined as an irreversible increase in volume or size. Growth invariably occurs as an increase in volume of individual cells. During the development of a whole plant growth is, of course, accompanied by cell division, but cell division per se does not cause enlargement.

Cellular differentiation is the transformation of apparently identical cells, arising from a common progenitor cell, into diverse cells with different biochemical, physiological and structural specializations. It results from differences in the manifestation of genetic potential in totipotent cells.

Morphogenesis refers to the origin of gross form and appearance of new organs. It encompasses growth and cellular differentiation but is a process of a higher order of magnitude which supercedes events occurring in single cells. Interactions among cells constitute a profound influence on the collective fates of individual cells in the developing plant body.

Certainly one of the most important unsolved problems in all of biology is the regulation of growth and development. The experimental approaches to this problem have been, and remain currently, many and varied. The knowledge, concepts and investigational tools and techniques of biochemistry-molecular biology, physiology, cytology and anatomy all are brought to bear. But whether biochemical investigations of cell-free systems or studies of origin of gross morphological features of whole plants, the overlying objectives are the same: to understand how one cell gives rise to millions of specialized cells all partitioned so precisely into interacting, interdependent tissues and organs and maintaining all the while such remarkable structural and functional unity as a living system.

An experimental procedure which is extensively employed in investigations of the regulation of plant growth and development is the culturing of excised plant parts and specific tissue explants in vitro on synthetic culture media. Tissue explants differ markedly in the success with which

95

they can be cultured _in vitro_, in their requirements for growth, and in the degree of morphogenesis, if any, they exhibit in culture. At one end of the gamut are numerous examples of tissues which grow very well on a simple medium containing only a few inorganic salts, a carbon and energy source and a few vitamins. Explants of meristems and of some tumor tissues, like those induced by the Crown Gall bacterium (_Agrobacterium tumefaciens_), are typical examples. At the opposite extreme are examples of tissue explants which so far have defied all attempts to grow them on defined media. In between the two extremes are numerous examples of explants from non-meristematic tissues which require for growth _in vitro_ an _auxin_ and a _cytokinin_. The most common result obtained with tissue explants cultured _in vitro_ is the proliferation of the explant into a mass of relatively undifferentiated cells, termed _callus_. But, as will be evident in further discussion, morphogenesis can be induced in at least some tissue cultures by appropriately regulating the chemical composition of the medium.

Professor Folke Skoog and his associates at the University of Wisconsin have made very important progress in the years since 1947 with explants of pith parenchyma from internodes of tobacco (_Nicotiana tabacum_ cv. Wisconsin No. 38) plants. Explants of this tobacco pith fail completely to grow on a basal medium unless supplied with an auxin such as _indoleacetic acid_ (IAA) (Fig. 9-1). On media containing auxin the pith explants exhibit cell enlargement but no cell division and no morphogenesis.

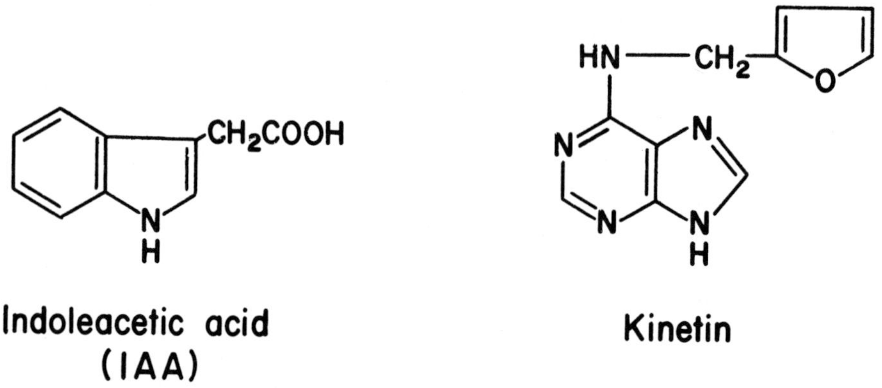

Indoleacetic acid
(IAA)

Kinetin

Fig. 9-1. Structures of indoleacetic acid, an auxin, and kinetin, a cytokinin.

Skoog and his associates discovered that if vascular tissue was placed in contact with the pith explants cultured _in vitro_ on a basal medium containing auxin, the tissue was stimulated to undergo cell division. A long search for the chemical stimulus supplied by the vascular tissue led ultimately to the discovery in Skoog's laboratory in 1954-55 of the _cytokinin-type plant hormones_. The first compound to be identified which could replace the cell division-stimulating influence of the vascular tissue on tobacco pith explants was <u>kinetin</u> (6-furfuryl-amino purine, $C_{10}H_9N_5O$) (Fig. 9-1). Kinetin was not isolated from tobacco tissues, and it may even be an artifact. However, the discovery of kinetin led to the subsequent discovery of several naturally occurring <u>cytokinins</u> from various microbial, plant and animal sources. The cyto-kinins are now firmly established as a major group of plant hormones which participate in the normal regulation of growth and development.

In a classic paper published in 1957, Skoog and Miller described a unique example of hormonal regulation of morphogenesis in tobacco pith explants. This important research beautifully documented a delicate and quantitative interaction between auxin and cytokinin in the control of bud and root formation from pith explants. They showed that with a par-ticular combination of concentrations of IAA and kinetin, the pith tissues grew as relatively undifferentiated callus. By varying the ratio of IAA:kinetin, however, they could successfully cause the explants to give rise to buds or to roots. Thus by varying the amounts of the two types of growth substances in the culture medium, morphogenesis in tobacco pith explants can be controlled to a remarkable degree. The implications for this kind of research for understanding normal, as well as abnormal or pathological, growth are great indeed.

The purpose of this exercise is to investigate the effects of IAA and kinetin on growth and morphogenesis in tobacco pith tissue <u>in vitro</u>.

Materials and Methods

At the laboratory period when the experiment is to be started, each team of students will be provided with 6 sterile, cotton-stoppered 125-ml Erlenmeyer flasks containing 50 ml each of culture medium; a sterile packet of essential glassware and utensils; and either a callus tissue culture or a vigorous tobacco (<u>Nicotiana tabacum</u>) plant (approximately 0.5 to 1.0 meter tall).

The basal culture medium[1] is that formulated by Linsmaier and Skoog

---

[1]Steps in preparation:
  a. Dissolve all the mineral salts in doubly distilled water and adjust the pH to 5.6 to 6.0 with 1 N NaOH.
  b. Dissolve the required amounts of <u>myo</u>-inositol and thiamine-HCl directly in minimal volumes of doubly distilled water.
  c. Add the required amount of kinetin to a minimal volume of 0.05 N NaOH and heat at 50 to 60° C until the kinetin dissolves.
  d. Dissolve the required amount of IAA in a minimal volume of 70 to 95% ethanol.
  e. Dissolve the required amount of sucrose in the mineral salts solution; add the solutions of <u>myo</u>-inositol, thiamine, kinetin and IAA; mix thoroughly.
  f. Heat the solution to boiling and add the required amount of agar.

(1965) and has the following composition:

Constituent	Concentration	Constituent	Concentration
Sucrose	30.0 g/liter	H_3BO_3	6.200 mg/liter
Agar	10.0 g/liter	KI	0.830 mg/liter
KNO_3	1900.0 mg/liter	Thiamine·HCl	0.400 mg/liter
NH_4NO_3	1650.0 mg/liter	$Na_2MoO_4·2H_2O$	0.250 mg/liter
$CaCl_2·2H_2O$	440.0 mg/liter	$CuSO_4·5H_2O$	0.025 mg/liter
$MgSO_4·7H_2O$	370.0 mg/liter	$CoCl_2·6H_2O$	0.025 mg/liter
KH_2PO_4	170.0 mg/liter	Indole-3-acetic	
myo-inositol	100.0 mg/liter	acid[3]	
NaFe-EDTA[2]	50.0 mg/liter	Kinetin[3]	
$MnSO_4·4H_2O$	22.3 mg/liter		
$ZnSO_4·4H_2O$	8.6 mg/liter		

The culture medium in each flask contains one of the following combinations of indoleacetic acid (IAA) and kinetin:

Flask No.	Conc. of IAA (mg/liter)	Conc. of kinetin (mg/liter)
1	0.00	0.00
2	0.00	0.20
3	2.00	0.00
4	2.00	0.20
5	2.00	0.02
6	0.02	2.00

All procedures involved in transplanting tissue into the culture flasks must be done with aseptic technique, that is, with utmost care to avoid contamination of the cultures with bacteria and fungi. Therefore, each team should perform all the critical operations either in a sterile transfer room or using a sterile transfer hood. Tissue explants must not be directly touched but rather must be handled with forceps or other utensils which have been sterilized in advance and dipped in 70% ethanol immediately before use. Tissues and utensils should not be laid on the working surface directly.

———————————

    g. Mix solution thoroughly.
    h. Pour 50 ml of the medium into each culture flask.
    i. Plug the flasks with cotton.
    j. Autoclave the flasks at 15 psi for 15 minutes.

[2]Linsmaier and Skoog (1965) specify 5 ml/liter of a stock solution containing 5.57 g $FeSO_4·7H_2O$ and 7.45 g Na_2EDTA per liter of doubly distilled water.

[3]IAA and kinetin levels are varied according to the type of growth desired. It is recommended that stock cultures be grown routinely on basal medium containing 2.0 mg/liter IAA and 0.2 mg/liter kinetin. On this medium, large uniform discs of firm white tissue of excellent quality for subculturing are formed.

Either of two alternative sources of tissue explants will be used, depending upon the length of time available to do the experiment and the preference of the instructor. If the experiment must be completed within 6 to 8 weeks, an actively growing tobacco callus culture should be used. Explants may be taken directly from a growing plant if the experiment can be continued for as long as 12 to 14 weeks.

Alternative 1 - Explants of callus. Open the cotton-stoppered culture flask, and with a sterile scalpel cut off small pieces (about 5 mm in diameter) from the callus. Using a sterile forceps or other utensil, transfer 3 callus explants to each of the 6 flasks. Terminate the experiment approximately 6 weeks later.

Alternative 2 - Explants from tobacco plant. Cut off the apical 20 to 30 cm of the shoot, and remove the leaves (including petioles) and the lateral and apical buds. Take the piece of stem to the sterile room or transfer hood. With a cotton swab dipped in 70% ethanol, thoroughly sterilize the surface of the stem. Cut off about 1 cm from each end with a sterile scalpel and discard these pieces. Cut the remaining piece into segments about 2 cm long. Holding each segment with a sterile forceps, punch out a cylinder of the central pith with a sterile cork borer (no. 3 or 4). Then cut from the cylinder 18 disks, each about 5 mm thick. Transfer 3 disks to each culture flask. Terminate the experiment approximately 8 to 10 weeks later.

Whichever alternative is followed, place the culture flasks in the designated controlled-environment facility, where they will be kept under constant light (100 to 200 ft-c) and temperature (about 28° C) for the duration of the experiment. Observe the cultures at each laboratory period. Record observations on the growth and morphogenesis of each culture, paying particular attention to such things as when growth is first apparent in each, evidence for the formation of shoots and roots, and other features. During the laboratory period when the experiment is to be terminated, write a detailed description of the condition of each culture, and make a detailed sketch of each flask and contents.

References

1. Fox, J. E. 1969. The cytokinins. Pp. 85-123 in: M. B. Wilkins, Ed. The Physiology of Plant Growth and Development. McGraw-Hill Book Company, New York.

2. Galston, A. W. and P. J. Davies. 1970. Control Mechanisms in Plant Development. Prentice-Hall, Englewood Cliffs, New Jersey.

3. Helgeson, J. P. 1968. The cytokinins. Science 161: 974-981.

4. Jablonski, J. R. and F. Skoog. 1954. Cell enlargement and cell division in excised tobacco pith tissue. Physiol. Plantarum 7: 16-24.

5. Leopold, A. C. and P. E. Kriedemann. 1975. Plant Growth and Development. 2nd Ed. McGraw-Hill Book Company, New York.

6. Letham, D. S. 1967. Chemistry and physiology of kinetin-like compounds. Ann. Rev. Plant Physiol. 18: 349-364.

7. Letham, D. S. 1969. Cytokinins and their relation to other phytohormones. BioScience 19: 309-316.

8. Linsmaier, E. M. and F. Skoog. 1965. Organic growth factor requirements of tobacco tissue cultures. Physiol. Plantarum 18: 100-127.

9. Miller, C. O., F. Skoog, M. N. von Saltza, and F. M. Strong. 1955. Kinetin, a cell division factor from deoxyribonucleic acid. J. Am. Chem. Soc. 77: 1392.

10. Moore, T. C. 1979. Biochemistry and Physiology of Plant Hormones. Springer-Verlag, New York.

11. Murashige, T. and F. Skoog. 1962. A revised medium for rapid growth and bioassays with tobacco tissue cultures. Physiol. Plantarum 15: 473-497.

12. Phillips, I. D. J. 1971. Introduction to the Biochemistry and Physiology of Plant Growth Hormones. McGraw-Hill Book Company, New York.

13. Salisbury, F. B. and C. W. Ross. 1978. Plant Physiology. 2nd Ed. Wadsworth Publishing Company, Belmont, California.

14. Skoog, F. and D. J. Armstrong. 1970. Cytokinins. Ann. Rev. Plant Physiol. 21: 359-384.

15. Skoog, F. and C. O. Miller, 1957. Chemical regulation of growth and organ formation in plant tissues cultured in vitro. Symp. Soc. Exp. Biol. 11: 118-131.

16. Skoog, F., F. M. Strong and C. O. Miller. 1965. Cytokinins. Science 148: 532-533.

17. Steward, F. C. 1968. Growth and Organization in Plants. Addison-Wesley Publishing Company, Reading, Massachusetts.

18. Steward, F. C. and A. D. Krikorian. 1971. Plants, Chemicals and Growth. Academic Press, New York.

19. Wareing, P. F. and I. D. J. Phillips. 1970. The Control of Growth and Differentiation in Plants. Pergamon Press, New York.

Special Materials and Equipment Required
Per Team of 3-8 Students

(1) Tobacco (_Nicotiana_ _tabacum_) plant or tobacco pith callus culture
(6) Cotton-plugged 125-ml Erlenmeyer flasks, each containing 50 ml of
    basal culture medium and one combination of concentrations of IAA
    and kinetin (see Materials and Methods)
(1) Controlled-environment facility
(1) Transfer hood or room
(1) Packet of sterile instruments: scalpel, cork borer (No. 3 or 4),
    forceps (6- or 8-inch length)
(Approximately 50 ml) 70% ethanol

Recommendations for Scheduling

This experiment should be performed by teams of 3 to 8 students.
All special materials should be prepared in advance and made available
to the students at the beginning of the laboratory period when the
experiment is to be started. Since the experiment extends for either
6 to 8 or 12 to 14 weeks, depending upon the alternative procedure
adopted, it is essential to start the tissue cultures early in the
term.

Duration: 6 to 8 or 12 to 14 weeks, approximately 2 hours of first
period, 15 minutes each succeeding period, and 1 hour final period.

Exercise 9

Effects of Auxin and Cytokinin on Morphogenesis in Callus Tissue

REPORT

Name _____ Section _____ Date _____

Periodic observations of cultures:

Flask No. 1 (0 mg/liter IAA; 0 mg/liter kinetin):

Flask No. 2 (0 mg/liter IAA; 0.2 mg/liter kinetin):

Flask No. 3 (2 mg/liter IAA; 0 mg/liter kinetin):

Flask No. 4 (2 mg/liter IAA; 0.2 mg/liter kinetin):

Flask No. 5 (2 mg/liter IAA; 0.02 mg/liter kinetin):

Flask No. 6 (0.02 mg/liter IAA; 2 mg/liter kinetin):

EFFECTS OF AUXIN AND CYTOKININ ON MORPHOGENESIS IN CALLUS TISSUE

REPORT

Name _____   Section _____   Date _____

Sketches of cultures at end of experiment:

Flask No. 1

Flask No. 4

Flask No. 2

Flask No. 5

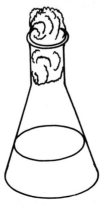

Flask No. 3

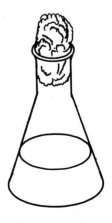

Flask No. 6

EXERCISE 9

EFFECTS OF AUXIN AND CYTOKININ ON MORPHOGENESIS IN CALLUS TISSUE

REPORT

Name _____ Section _____ Date _____

Questions

1. Did growth or morphogenesis of the explants occur in the absence of either IAA or kinetin or both substances? Explain.

2. How did the different ratios of IAA and kinetin affect morphogenesis in the explants? Describe in detail.

3. If buds and roots were observed to form, did they originate directly from the original explant or from callus tissue which proliferated from the explant? Describe explicitly.

4. How could the results of this experiment be quantified?

Exercise 10

Role of Phytochrome in the Germination of Light-sensitive Lettuce Seeds

Introduction

Phytochrome is a water-soluble pigment which functions as the photo-receptor for a wide variety of plant responses to light, indeed all the so-called "red, far-red reversible photoreactions" (Table 10-I). In participating in these diverse responses, phytochrome influences the growth, development and behavior of seed plants literally throughout their ontogeny. Furthermore, phytochrome is by no means restricted to seed plants. It has definitely been identified in certain liverworts and green algae, for example, and there is evidence for the speculation that it probably occurs in all green plants.

The definitive discovery of phytochrome was made by a group of in-vestigators (H. A. Borthwick, S. B. Hendricks, H. W. Siegelman and K. H. Norris) at the Agricultural Research Center of the U. S. Department of Agriculture at Beltsville, Maryland in 1959. Since that time much has been learned about the biochemistry of the pigment and the ways by which it affects plants. Chemically, phytochrome is a water-soluble chromo-protein, which appears to be at least partially localized in or associated with one or more membrane fractions of cells, including the plasmalemma. The molecular weight of the protein moiety of native phytochrome has not been precisely determined. The molecular weight estimates that have been made with highly purified preparations of phytochrome vary from as little as 36,000 to as much as 180,000. The most convincing evidence available currently is for a monomer molecular weight of approximately 120,000. The chromophore is an open-chain tetrapyrrole bilitriene pigment similar to phycocyanin. Available evidence suggests that there is one molecule of chromophore per molecule of protein.

Phytochrome exists in two spectrophotometrically readily distinguish-able and photointerconvertible forms, P_r and P_{fr}, and several transient intermediate forms. P_r is blue and exhibits maximum absorption at about 660 nm, in the red portion of the visible spectrum. P_{fr} is blue-green and absorbs maximally at about 730 nm, in the far-red (Fig. 10-1). Upon irradiation with monochromatic red light, P_r is converted to P_{fr}; upon irradiation with far-red light, P_{fr} is transformed to P_r. With narrow-band far-red light, P_{fr} can be nearly completely transformed to P_r. However, because of the overlapping of the absorption spectra in the region of 600 to 700 nm, the most effective attainable band of red light transforms only about 80% of P_r to P_{fr}. Both P_r and P_{fr} also exhibit significant absorption in the blue region of the spectrum, the absorption maxima being at approximately 370 and 400 nm, respectively (Fig. 10-1).

Table 10-I. Classification of Phytochrome-mediated Photoresponses

I.   Photoperiodic, photomorphogenetic responses.
     A. Flowering of photoperiodically sensitive species.
     B. Vegetative photoperiodic responses.
         1. Onset of winter bud dormancy in woody perennials.
         2. Breaking of winter bud dormancy in woody perennials.
         3. Leaf abscission in deciduous woody plants.
         4. Cessation of cambial activity in woody perennials.
         5. Germination of some kinds of seeds.
         6. Tuber formation in some plants.
         7. Bulb formation in some plants.

II. Non-photoperiodic, photomorphogenetic responses.
     A. De-etiolation responses of angiosperm seedlings.
         1. Stimulation of leaf-blade expansion in dicots, leaf
             unrolling in monocots.
         2. Inhibition of stem elongation.
         3. Opening of plumular or hypocotyl hooks of dicots.
         4. Development of chloroplasts--see III-A-1.
     B. Germination of some kinds of seeds (e.g. Grand Rapids lettuce).
     C. Phototaxis of the chloroplasts of Mougeotia and other algae.
     D. Elongation of fern rhizoids.
     E. Nyctinastic leaf movements in Mimosa pudica and related legumes.
     F. Modification of phototropism and geotropism (geotropism is
         stimulated by previous exposure to red light; first positive
         and negative phototropic curvatures are reduced by previous
         exposure to red light, second positive curvature stimulated).

III. Non-morphogenetic photoresponses.
     A. Stimulation of synthesis of enzymes and other proteins in
         etiolated leaves upon illumination.
         1. Chloroplastic proteins.
             a. Structural proteins.
             b. Ribulose-1,5-diphosphate carboxylase.
             c. NADP-linked glyceraldehyde-3-phosphate dehydrogenase.
         2. Mitochondrial enzymes--glutamic dehydrogenase.
         3. Soluble (cytoplasmic) enzymes.
             a. Glycolic acid oxidase.
             b. Glucose-6-phosphate dehydrogenase.
             c. NAD-linked glyceraldehyde-3-phosphate dehydrogenase.
     B. Synthesis of anthocyanins and other flavonoids in a variety of
         tissues, e.g. induction of phenylalanine ammonia lyase
         (phenylalanine deaminase) in epidermis of etiolated white
         mustard (Sinapis alba L.) seedlings.
     C. Stimulation of gibberellin biosynthesis.

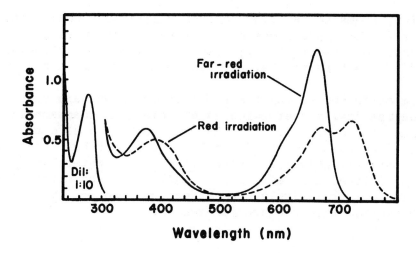

Fig. 10-1. Absorption spectra of a solution of oat phytochrome following irradiation with red and far-red light. The solid line thus represents the absorption spectrum for P_r. (Redrawn from Siegelman and Butler, 1965.)

Basically four major transformation reactions have been described for phytochrome (Fig. 10-2). Reactions I and II, discussed previously, both are first-order photochemical reactions. Reaction IV, a destruction

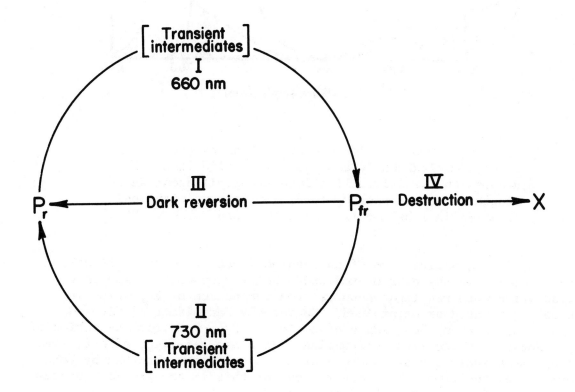

Fig. 10-2. Phytochrome transformation reactions.

or decay reaction, seems to be ubiquitous in all tissues containing phytochrome. It results in a decrease in total photoreversible phytochrome. Dark reversion of P_{fr} to P_r, reaction III, has been demonstrated in various dicots but as yet has not been demonstrated in monocots.

The quantum efficiencies for the two photoreactions are such that reaction I is energetically over twice as efficient as reaction II (Fig. 10-3). The action of blue light is very much less effective than red and far-red light in reactions I and II. Red light has been reported to be 100 times more effective than blue, in vivo in converting P_r, and far-red light is 25 times more efficient than blue in converting P_{fr}. In vitro, blue light is relatively more effective than in vivo but still markedly less effective than red and far-red light.

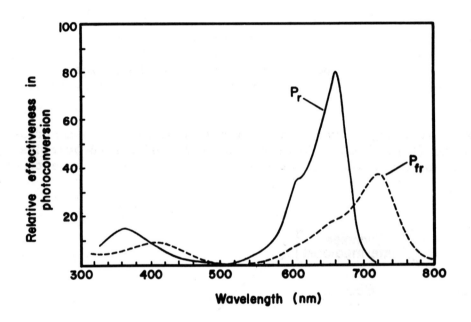

Fig. 10-3. Action spectra of the photochemical transformations of P_r and P_{fr}. The solid line thus denotes the relative effectiveness of different wavelengths in converting P_r to P_{fr}. (Redrawn with modification in labeling from Siegelman and Butler, 1965.)

P_{fr} is the active form of phytochrome, and P_r is metabolically inert. Thus, in phytochrome-mediated photoresponses, a response to irradiation with red light results from the action of P_{fr} which is thereby generated or maintained. Conversely, an effect of far-red light results from the absence of action of P_{fr}—not from any action of P_r. Notably, white light or sunlight has a net red-light effect, even though such radiation of course contains wavelengths absorbed by both forms of phytochrome. The main reason for this is the greater quantum efficiency of reaction I than reaction II.

The germination of many kinds of seeds is strongly influenced by light. In some species, e.g. the Grand Rapids variety of lettuce (Lactuca sativa), germination is promoted and may be absolutely dependent upon red light. The effect of an irradiation with red light is nullified by immediate subsequent exposure to far-red light. Seeds of other species are promoted by red light but do not have an obligatory light requirement. Germination of seeds of still other species is inhibited or prevented by light, and finally the germination of some kinds of seeds is unaffected by light.

Two major hypotheses have been advanced regarding the mechanism of action of phytochrome (specifically P_{fr}). One hypothesis regards P_{fr} as acting through a differential gene activation mechanism. The other considers regulation of membrane permeability as the mechanism of action. There is considerable evidence which supports both of these hypotheses, although the weight of evidence in the majority of recent investigations is in support of the latter. It may be that the pigment has more than one fundamental site and mechanism of action. Additional research is required to settle the question.

The purpose of this experiment is to investigate the effects of red and far-red light on the germination of Grand Rapids lettuce seeds (actually achenes).

Materials and Methods

Put several hundred Grand Rapids lettuce seeds[1] into a petri dish or other suitable germinator containing a double layer of filter paper moistened with distilled water 4 to 16 hours prior to the time of light treatment. Cover the container with aluminum foil to make it absolutely light-tight.

Each team of students prepare 4 10-cm petri dishes as follows: place a disk of 9-cm filter paper in the bottom of each dish and moisten the paper with 3 to 5 ml of distilled water. Obtain 4 sheets of aluminum foil large enough to make a light-tight wrap on a single dish. Take the petri dishes, a forceps, and a marking pencil to the darkroom, where the lighting apparatus is assembled.

While working under dim blue light,[2] transfer (with forceps) 50 hydrated seeds to each of the 4 petri dishes. Place the lid on one dish immediately, wrap the dish with foil, and label it "hydrated dark control." Now, with the lids left off, expose all 3 remaining dishes to 2 minutes

---

[1] Obtain within one year of harvest from Carolina Biological Supply Company, Burlington, North Carolina 27215, or other known reliable source.

[2] A 25-watt incandescent lamp mounted in a gooseneck desk lamp and covered with 4 layers of DuPont blue cellophane.

(or 4 minutes)[3] of red light.[4] Then remove, wrap, and label one dish immediately. Expose the remaining 2 dishes immediately to 4 minutes (or 8 minutes) irradiation with far-red light.[5] Quickly remove, wrap, and label one of the dishes. Now re-expose the sole remaining dish to 2 minutes (or 4 minutes) of red light; then wrap and label it immediately. Finally, upon return to the laboratory, place 50 dry seeds in another (5th) petri dish containing a disk of filter paper. Moisten the paper with distilled water; then promptly wrap and label this dish as "non-hydrated dark control." Place all the dishes in a dark cabinet or drawer. Determine the percentage germination in each group of seeds 72 to 96 hours later. Regard as germinated only those seeds which exhibit unequivocal protrusion of the radicle. Present the collective class results in a table.

References

1. Bellini, E. and W. S. Hillman. 1971. Red and far red effects on phenylalanine ammonia-lyase in Raphanus and Sinapis seedlings do not correlate with phytochrome spectrophotometry. Plant Physiol. 47: 668-671.

2. Bewley, J. D., M. Black, and M. Negbi. 1967. Immediate action of phytochrome in light-stimulated lettuce seeds. Nature 215: 648-649.

3. Borthwick, H. A. and S. B. Hendricks. 1960. Photoperiodism in plants. Science 132: 1223-1228.

4. Borthwick, H. A., S. B. Hendricks, M. W. Parker, E. H. Toole, and V. K. Toole. 1952. A reversible photoreaction controlling seed germination. Proc. Nat. Acad. Sci. 38: 662-666.

---

[3]Two-minute irradiations with red light, alternated with 4-minute irradiations with far-red light is a satisfactory regimen for the conditions employed; alternatively, exposures of 4 minutes red and 8 minutes far-red may be used successfully.

[4]As a red-light source, the author routinely uses a 100-watt incandescent lamp mounted in a 10-inch photographic reflector, on which is mounted with a taped cardboard mask a Corning red glass filter, No. M2398, 16.5 x 16.5 cm. The reflector is attached to a ringstand with the filter 12 to 18 inches above the seeds.

[5]The author uses a 100-watt incandescent lamp mounted in a 10-inch photographic reflector and covered with a Corning far-red glass filter, No. 2707, 16.5 x 16.5 cm, plus one layer of DuPont blue cellophane. The filter is elevated 4 to 6 inches above the seeds.

5. Borthwick, H. A., S. B. Hendricks, E. H. Toole, and V. K. Toole.
    1954. Action of light in lettuce seed germination. Botan.
    Gaz. 115: 205-225.

6. Briggs, W. R., W. D. Zollinger, and B. B. Platz. 1968. Some
    properties of phytochrome from dark-grown oat seedlings
    (Avena sativa L.). Plant Physiol. 43: 1239-1243.

7. Butler, W. L. and R. J. Downs. 1960. Light and plant develop-
    ment. Scient. Am. 203 (No. 6, Dec.): 56-63.

8. Butler, W. L., K. H. Norris, H. W. Siegelman, and S. B. Hendricks.
    1959. Detection, assay, and preliminary purification of the
    pigment controlling photoresponsive development of plants.
    Proc. Nat. Acad. Sci. 45: 1703-1708.

9. Correll, D. L. and J. L. Edwards. 1970. The aggregation states
    of phytochrome from etiolated rye and oat seedlings. Plant
    Physiol. 45: 81-85.

10. Cross, D. R., H. Linschitz, V. Kasche, and J. Tenenbaum. 1968.
    Low-temperature studies on phytochrome: Light and dark
    reactions in the red to far-red transformation and new
    intermediate forms of phytochrome. Proc. Nat. Acad. Sci.
    61: 1095-1101.

11. Filner, B. and A. O. Klein. 1968. Changes in enzymatic
    activities in etiolated bean seedling leaves after a brief
    illumination. Plant Physiol. 43: 1587-1596.

12. Fondeville, J. C., H. A. Borthwick, and S. B. Hendricks. 1966.
    Leaflet movement of Mimosa pudica L. indicative of phyto-
    chrome action. Planta 69: 357-364.

13. Galston, A. W. 1968. Microspectrophotometric evidence for
    phytochrome in plant nuclei. Proc. Nat. Acad. Sci.
    61: 454-460.

14. Galston, A. W. and P. J. Davies. 1970. Control Mechanisms in
    Plant Development. Prentice-Hall, Englewood Cliffs,
    New Jersey.

15. Haupt, W. 1965. Perception of environmental stimuli orienting
    growth and movement in lower plants. Ann. Rev. Plant
    Physiol. 16: 267-290.

16. Hendricks, S. B. 1963. Metabolic control of timing. Science
    141: 21-27.

17. Hendricks, S. B. 1968. How light interacts with living matter.
    Scient. Am. 219 (No. 3, Sept.): 174-186.

18. Hendricks, S. B. and H. A. Borthwick. 1967. The function of phytochrome in regulation of plant growth. Proc. Nat. Acad. Sci. 58: 2125-2130.

19. Hillman, W. S. 1967. The physiology of phytochrome. Ann. Rev. Plant Physiol. 18: 301-324.

20. Hillman, W. S. and W. L. Koukkari. 1967. Phytochrome effects in the nyctinastic leaf movements of Albizzia julibrissin and some other legumes. Plant Physiol. 42: 1413-1418.

21. Ikuma, H. and K. V. Thimann. 1963. The role of the seed coats in germination of photosensitive lettuce seeds. Plant and Cell Physiol. 4: 169-185.

22. Ikuma, H. and K. V. Thimann. 1964. Analysis of germination of lettuce seed by means of temperature and anaerobiosis. Plant Physiol. 39: 756-767.

23. Jaffe, M. J. 1968. Phytochrome-mediated bioelectric potentials in mung bean seedlings. Science 162: 1016-1017.

24. Jaffe, M. J. 1970. Evidence for the regulation of phytochrome-mediated processes in bean roots by the neurohumor, acetylcholine. Plant Physiol. 46: 768-777.

25. Jaffe, M. J. and A. W. Galston. 1967. Phytochrome control of rapid nyctinastic movements and membrane permeability in Albizzia julibrissin. Planta 77: 135-141.

26. Lange, H., I. Bienger, and H. Mohr. 1967. New evidence in favour of the hypothesis of differential gene activation by phytochrome 730. Planta 76: 359-366.

27. Lange, H., W. Shropshire, Jr., and H. Mohr. 1971. An analysis of phytochrome-mediated anthocyanin synthesis. Plant Physiol. 47: 649-655.

28. Linschitz, H., V. Kasche, W. L. Butler, and H. W. Siegelman. 1966. The kinetics of phytochrome conversion. J. Biol. Chem. 241: 3395-3403.

29. Mancinelli, A. L., H. A. Borthwick, and S. B. Hendricks. 1966. Phytochrome control of tomato seed germination. Botan. Gaz. 127: 1-5.

30. Mayber, A. M. and A. Poljakoff-Mayber. 1964. The Germination of Seeds. Pergamon Press, Oxford.

31. Mohr, H. 1966. Differential gene activation as a mode of action of phytochrome 730. Photochem. Photobiol. 5: 469-483.

32. Moore, T. C. 1979. Biochemistry and Physiology of Plant Hormones. Springer-Verlag, New York.

33. Pratt, L. H. and W. R. Briggs. 1966. Photochemical and non-photochemical reactions of phytochrome in vivo. Plant Physiol. 41: 467-474.

34. Reid, D. M., J. B. Clements, and D. J. Carr. 1968. Red light induction of gibberellin synthesis in leaves. Nature 217: 580-582.

35. Rissland, I. and H. Mohr. 1967. Phytochrome-mediated enzyme formation (phenylalanine deaminase) as a rapid process. Planta 77: 239-249.

36. Salisbury, F. B. and C. W. Ross. 1978. Plant Physiology. 2nd Ed. Wadsworth Publishing Company, Belmont, California.

37. Satter, R. L., P. B. Applewhite, and A. W. Galston. 1972. Phytochrome-controlled nyctinasty in Albizzia julibrissin. V. Evidence against acetylcholine participation. Plant Physiol. 50: 523-525.

38. Schopfer, P. and H. Mohr. 1972. Phytochrome-mediated induction of phenylalanine ammonia-lyase in mustard seedlings. A contribution to eliminate some misconceptions. Plant Physiol. 49: 8-10.

39. Siegelman, H. W. and W. L. Butler. 1965. Properties of phytochrome. Ann. Rev. Plant Physiol. 16: 383-392.

Special Materials and Equipment Required
Per Team of 3-4 Students

(200) Hydrated Grand Rapids lettuce (Lactuca sativa) seeds in a light-tight container (seeds obtainable, e.g., from Carolina Biological Supply Company, Burlington, North Carolina 27215)
(50) Non-hydrated Grand Rapids lettuce seeds
(1) Irradiation apparatus:
    (1) Red light source
    (1) Far-red light source
    (1) Blue safelight
(1) Darkroom
(5) 10-cm petri dishes
(5) Disks of 9-cm filter paper
(1) Forceps
(1) 100-ml beaker containing 50 ml of 70 or 95% ethanol
(1) Polyethylene washing bottle filled with distilled water
(Approximately 6 linear feet) Aluminum foil

Recommendations for Scheduling

Hydrated seeds, in a light-tight container, should be available at the beginning of the laboratory period when the experiment is to be started. Students should work in teams of 3 to 4 members. One or two students each from all teams may go to the darkroom at one time to plate out the hydrated seeds and administer the light treatments. Alternatively, whole teams may individually take turns going to the darkroom.

NOTE: Occasionally batches of Grand Rapids lettuce seeds obtained from commercial sources behave contrary to expectations and germinate very well in darkness, that is, exhibit no light requirement whatsoever. This can be perplexing to unsuspecting instructors, and it would be interesting to know why some seeds behave that way. It may be the result of simply air-drying freshly harvested seed in the light, but this is only speculation. Whatever the explanation, such seeds still can be used successfully in this experiment by irradiating all the hydrated seeds with far-red light immediately prior to the other prescribed irradiations. Then only the "non-hydrated dark controls" would behave contrary to expectation. This problem is encountered frequently enough to make it advisable to check each new batch of Grand Rapids lettuce seeds before starting an actual experiment.

Duration: 3 to 4 days, parts of 2 periods; approximately 30 to 45 minutes each period.

ROLE OF PHYTOCHROME IN THE GERMINATION OF LIGHT-SENSITIVE LETTUCE SEEDS

REPORT

Name _____ Section _____ Date _____

Treatment	% Germination				
	Group 1	Group 2	Group 3	Group 4	Mean
Non-hydrated dark control	_____	_____	_____	_____	_____
Hydrated dark control	_____	_____	_____	_____	_____
R^1	_____	_____	_____	_____	_____
$R + FR^1$	_____	_____	_____	_____	_____
$R + FR + R^1$	_____	_____	_____	_____	_____

[1]Exposure times: red, _____ minutes; far-red, _____ minutes.

EXERCISE 10

ROLE OF PHYTOCHROME IN THE GERMINATION OF LIGHT-SENSITIVE LETTUCE SEEDS

REPORT

Name _____ Section _____ Date _____

Questions

1. Is there any evidence from this experiment that blue light affected germination? Explain.

2. Explain why the exposure time for far-red irradiation was twice that for red irradiation.

3. To what extent were the red and far-red light treatments mutually reversible?

4. What is a possible explanation for the observed action of red light on the germination of Grand Rapids lettuce seeds?

ROLE OF PHYTOCHROME IN THE GERMINATION OF
LIGHT-SENSITIVE LETTUCE SEEDS

REPORT

Name _____ Section _____ Date _____

5. Describe the survival value, if any, that might be associated with a light requirement for seed germination.

6. Describe particular chemical and temperature treatments which have been shown by others to promote the germination of Grand Rapids lettuce seeds in darkness, and discuss briefly whether these treatments act in fundamentally the same way as red light.

7. When irradiation sequences involving multiple alternating irradiations with red and far-red light have been used, it has been observed that germination percentages following terminal irradiations with far-red light increase steadily as the number of alternating red-far-red treatments increases. Explain this result.

Exercise 11

Dormancy of Seeds of White Ash (Fraxinus americana)

Introduction

Seeds of many species of angiosperms and gymnosperms will not germinate immediately after ripening even under conditions of moisture, temperature and oxygen tension generally favorable for growth. Such dormancy, like dormancy of buds, is of obvious survival value to plants, since it tends to restrict germination to environmental circumstances suitable for seedling establishment and survival. Different seeds in one seed crop often vary in their degree of dormancy, causing germination of the crop to occur over an extended period of time and thus increasing the probability that at least part of them will germinate under conditions favorable for survival.

Dormancy results from particular conditions prevailing internally in the seed. Several causes of dormancy have been recognized, and seeds of numerous species exhibit two or more of them simultaneously. Among the many causes of seed dormancy which have been described are: (1) impermeability of seed coats to water and gases; (2) immaturity of the embryo; (3) need for "after-ripening" in dry storage; (4) mechanical resistance of seed coats; (5) presence of inhibitors found either in the seed coats, dry accessory structures, or, in the case of seeds contained in fleshy fruits, in the tissues surrounding the seeds; (6) special requirement for light or its absence; and (7) requirement for chilling in the hydrated condition.

The most common cause of seed dormancy in woody plants native to temperate regions is a requirement for chilling of the hydrated seeds. Such seeds commonly germinate promptly and uniformly only after they have become hydrated and exposed to low temperature (0 to 5° C) for a period of a few to several weeks. When these conditions are supplied artificially, the practice is referred to as stratification. It is this type of dormancy with which this exercise is concerned.

Dormancy of seeds which have a chilling requirement appears to be regulated by two types of growth substances with opposing actions--growth inhibitors and growth-promoting hormones. There may be several to many different inhibitors causally involved in regulation of dormancy of seeds of different species. However, the inhibitor of primary, if not exclusive, importance in seeds of at least several species has been identified as abscisic acid (ABA) (Fig. 11-1). Gibberellins (GAs) (Fig. 11-1) are the growth-promoting substances which have been most directly implicated in the regulation of seed dormancy.

121

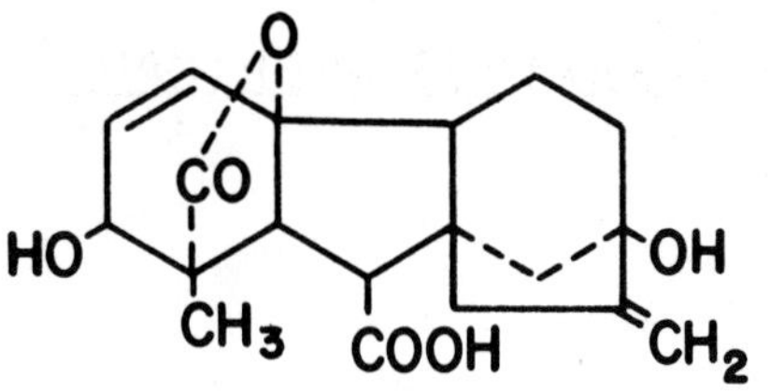

Gibberellin A₃ (GA₃)   **(S) - Abscisic acid [(S) - ABA]**

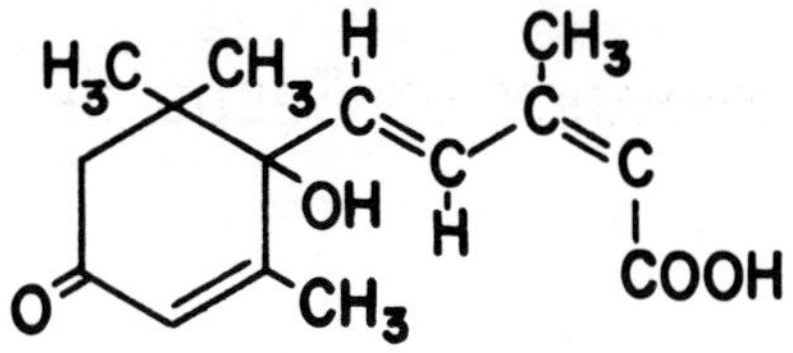

Fig. 11-1. Structures of gibberellin A$_3$ and (S)-abscisic acid.

The effects of stratification on levels of endogenous growth-promoting hormones (particularly GAs) and inhibitors, and the interactions between promoting and inhibitory substances, have been investigated intensively in recent years. Seeds of certain ash (Fraxinus) species and of filbert or hazelnut (Corylus avellana) have received particular attention. Several tentative generalizations have emerged from these studies. One is that treatment with GA of several kinds of dormant seeds which normally require chilling causes the seeds to germinate in the absence of low-temperature treatment. A second generalization is that the endogenous GA content of at least some such seeds increases during chilling or at germination temperatures subsequent to chilling. Thirdly, GA opposes the action of ABA on a number of physiological processes, including germination of seeds with a chilling requirement. A fourth generalization is that in certain kinds of seeds the ABA content has been shown to decrease markedly during chilling.

Thus it appears that dormancy of seeds with a chilling requirement may be regulated by a balance between specific growth-promoting hormones and growth inhibitors. The dormant state is correlated with a relatively high inhibitor : promoter ratio. Chilling causes a shift in this balance through a depletion of inhibitor, increase in promoter, or, perhaps in some cases, both of these processes.

The purpose of this experiment is to investigate the effects of stratification and exogenous growth-regulating chemicals on embryo dormancy and germination in seeds of White Ash (Fraxinus americana), which require chilling, and to compare the behavior of excised White Ash embryos with excised embryos of Flowering Ash (F. ornus) seeds which are non-dormant.

122

Materials and Methods[1]

Each team obtain from the bulk supply approximately 150 samaras (or naked seeds from which wings and pericarps have been removed) of White Ash (Fraxinus americana)[2] and about 50 samaras or seeds of Flowering Ash (F. ornus). If samaras are supplied, carefully remove the wing and pericarp by hand from each seed.

Surface-disinfect 100 F. americana seeds by placing them in a rubber-stoppered Erlenmeyer flask and washing them with vigorous agitation for several minutes in a solution of 0.1% sodium hypochlorite (approximately 2% household Clorox or Purex in sterile distilled water), and rinsing 5 times with sterile distilled water. Next, plant the seeds deeply in a large beaker (or polyethylene bag) of moist, sterile vermiculite. Cover the beaker with aluminum foil, cellophane, or Parafilm (or seal the bag), and set the container in a refrigerator or cold room at approximately 5° C for 60 to 90 days. (Alternatively the seeds can be stratified on moist filter paper in petri dishes.) Similarly prepare and stratify 25 seeds of F. ornus. Check the containers every two weeks or so; add sterile distilled water if the vermiculite is becoming dry.

On the day the stratification is concluded, remove the stratified seeds from the containers. Excise the embryos from the stratified seeds by carefully slicing open the endosperm with a sterile scalpel or razor blade, avoiding cutting the embryo, and remove the embryo with a sterile forceps. Place 20 excised embryos in each of 5 labeled, sterile 10-cm petri dishes each containing a sterile 9-cm disk of Whatman No. 1 filter paper. Also excise the embryos from 40 non-stratified F. americana seeds and 20 non-stratified F. ornus seeds which have been previously disinfected (as described above) and hydrated with distilled water under sterile conditions (seeds should not be submerged) for a preceding 24 hours. Place 20 of these embryos in each of 3 labeled, sterile petri dishes. Then to each petri dish add 3 ml of sterile 0.01 M potassium phosphate buffer (pH 6.0) or sterile solution of growth substance as indicated in the following table. All solutions are prepared with the phosphate buffer.

Label all the petri dishes for ready identification. Place them on a laboratory bench where they will be exposed to light or in a controlled-environment under constant light (approximately 50 to 250 ft-c) and a temperature of about 22° C. Check the embryos at each laboratory period.

---

[1]Adopted from: Sondheimer, E. and E. C. Galson. 1966. Effects of abscisin II and other plant growth substances on germination of seeds with stratification requirements. Plant Physiol. 41: 1397-1398; and Sondheimer, E., D. S. Tzou, and E. C. Galson. 1968. Abscisic acid levels and seed dormancy. Plant Physiol. 43: 1443-1447.

[2]Seeds of filbert (Corylus avellana) can be substituted for seeds of Fraxinus americana. Refer to papers by J. W. Bradbeer and associates for specific details of procedure. Filbert seeds are available from F. W. Schumacher Company, Sandwich, Massachusetts 02563.

Table 11-I. Treatments of F. americana and F. ornus Embryos

Group	Species	Stratification	Solution
1	F. americana	–	Buffer only
2	F. americana	+	Buffer only
3	F. americana	–	10 μM GA$_3$
4	F. americana	+	10 μM Amo-1618
5	F. americana	+	10 μM ABA
6	F. americana	+	10 μM ABA + 10 μM GA$_3$
7	F. ornus	–	Buffer only
8	F. ornus	+	Buffer only

Add additional water or chemical solution if necessary. Count the number of embryos which have germinated in each group 10 to 14 days after planting them in the petri dishes. Regard as germinated only those embryos with obvious geotropic curvature of the radicle. Record the germination percentages in a table.

References

1. Addicott, F. T. and J. L. Lyon. 1969. Physiology of abscisic acid and related substances. Ann. Rev. Plant Physiol. 20: 139-164.

2. Amen, R. D. 1968. A model of seed dormancy. Botan. Rev. 34: 1-32.

3. Bradbeer, J. W. 1968. Studies in seed dormancy. IV. The role of endogenous inhibitors and gibberellin in the dormancy and germination of Corylus avellana L. seeds. Planta 78: 266-276.

4. Bradbeer, J. W. and N. J. Pinfield. 1967. Studies in seed dormancy. III. The effects of gibberellin on dormant seeds of Corylus avellana L. New Phytol. 66: 515-523.

5. Frankland, B. and P. F. Wareing. 1962. Changes in endogenous gibberellins, in relation to chilling of dormant seeds. Nature 194: 313-314.

6. Frankland, B. and P. F. Wareing. 1966. Hormonal regulation of seed dormancy in hazel (Corylus avellana L.) and beech (Fagus sylvatica). J. Exp. Botany 17: 596-611.

7. Galston, A. W. and P. J. Davies. 1970. Control Mechanisms in Plant Development. Prentice-Hall, Englewood Cliffs, New Jersey.

8. Jarvis, B. C., B. Frankland, and J. H. Cherry. 1968. Increased DNA template and RNA polymerase associated with the breaking of seed dormancy. Plant Physiol. 43: 1734-1736.

9. Leopold, A. C. and P. E. Kriedemann. 1975. Plant Growth and Development. 2nd Ed. McGraw-Hill Book Company, New York.

10. Lipe, W. N. and J. C. Crane. 1966. Dormancy regulation in peach seeds. Science 153: 541-542.

11. Moore, T. C. 1979. Biochemistry and Physiology of Plant Hormones. Springer-Verlag, New York.

12. Phillips, I. D. J. 1971. Introduction to the Biochemistry and Physiology of Plant Growth Hormones. McGraw-Hill Book Company, New York.

13. Ross, J. D. and J. W. Bradbeer. 1968. Concentrations of gibberellin in chilled hazel seeds. Nature 220: 85-86.

14. Ross, J. D. and J. W. Bradbeer. 1971a. Studies in seed dormancy. V. The content of endogenous gibberellins in seeds of Corylus avellana L. Planta 100: 288-302.

15. Ross, J. D. and J. W. Bradbeer. 1971b. Studies in seed dormancy. VI. The effects of growth retardants on the gibberellin content and germination of chilled seeds of Corylus avellana L. Planta 100: 303-308.

16. Salisbury, F. B. and C. W. Ross. 1978. Plant Physiology. 2nd Ed. Wadsworth Publishing Company, Belmont, California.

17. Sondheimer, E. and E. C. Galson. 1966. Effects of abscisin II and other plant growth substances on germination of seeds with stratification requirements. Plant Physiol. 41: 1397-1398.

18. Sondheimer, E., D. S. Tzou, and E. C. Galson. 1968. Abscisic acid levels and seed dormancy. Plant Physiol. 43: 1443-1447.

19. Steward, F. C. and A. D. Krikorian. 1971. Plants, Chemicals and Growth. Academic Press, New York.

20. Thomas, T. H., P. F. Wareing, and P. M. Robinson. 1965. Action of the sycamore 'dormin' as a gibberellin antagonist. Nature 205: 1270-1272.

21. Timson, I. 1965. New method of recording germination data. Nature 207: 216.

22. Villiers, T. A. and P. F. Wareing. 1965. The possible role of low temperature in breaking the dormancy of seeds of Fraxinus excelsior L. J. Exp. Botany 16: 519-531.

23. Wareing, P. F. 1969. Germination and dormancy. Pp. 605-644
        in: M. B. Wilkins, Ed. The Physiology of Plant Growth and
        Development. McGraw-Hill Publishing Company, Ltd., London.

24. Wareing, P. F. and I. D. J. Phillips. 1970. The Control of
        Growth and Differentiation in Plants. Pergamon Press,
        New York.

Special Materials and Equipment Required
Per Team of 3-8 Students

(150) Samaras or naked seeds of White Ash (Fraxinus americana) (Seeds can
     be purchased from F. W. Schumacher Company, Sandwich, Massachusetts
     02563, or Herbst Brothers Seedsmen, Incorporated, Brewster, New York
     10509)
(50) Samaras or naked seeds of Flowering Ash (Fraxinus ornus) (Seeds
     sometimes can be purchased from F. W. Schumacher Company, Sandwich,
     Massachusetts 02563)
(8) Sterile 10-cm petri dishes each containing a sterile 9-cm disk of
     Whatman No. 1 filter paper
(Approximately 100 ml) 0.1% sodium hypochlorite solution
(Approximately 1 liter) Sterile distilled water
(1) 800-ml or 1-liter beaker filled with sterile vermiculite
(1) 250-ml beaker filled with sterile vermiculite
(Approximately 50 ml) 0.01 M potassium phosphate buffer (pH 6.0) (sterile)
(Approximately 50 ml each) Following solutions (sterile) prepared with
     0.01 M potassium phosphate buffer (pH 6.0):
     10 μM GA_3
     10 μM Amo-1618
     10 μM ABA
     10 μM ABA + 10 μM GA_3
(1) Cold room or refrigerator
(1) Controlled-environment facility

Recommendations for Scheduling

Since this experiment extends for a period of 70 to 100 days, it is
imperative that the stratification be started very early in the term. In
the case of one-quarter courses it may be desirable or necessary for the
instructional staff to start the stratification even before the course
begins. Students should work in teams of 3 to 8 members.

Duration: 70 to 100 days; approximately 45 minutes first period,
periodic brief checks thereafter until germination tests are started.

Exercise 11

Dormancy of Seeds of White Ash (Fraxinus americana)

REPORT

Name _____ Section _____ Date _____

Germination data:

| Group | Species | Treatment | | % Germination |
		Stratification	Chemical	
1	F. americana	–	Buffer	_____
2	F. americana	+	Buffer	_____
3	F. americana	–	GA	_____
4	F. americana	+	Amo–1618	_____
5	F. americana	+	ABA	_____
6	F. americana	+	ABA + GA	_____
7	F. ornus	–	Buffer	_____
8	F. ornus	+	Buffer	_____

Observations:

EXERCISE 11

DORMANCY OF SEEDS OF WHITE ASH (FRAXINUS AMERICANA)

REPORT

Name _____ Section _____ Date _____

Questions

1. Were the seeds of either or both ash species actually dormant at the start of the experiment? Explain.

2. For what reason were the germination tests performed with excised embryos instead of whole seeds?

3. What were the comparative effects of stratification on seeds of Fraxinus americana and F. ornus?

4. Describe the effect of exogenous GA on F. americana embryos, and discuss briefly how this effect relates to the natural role of endogenous GA in controlling dormancy in seeds which have a chilling requirement.

5. What effect, if any, did Amo-1618, a known inhibitor of GA biosynthesis, have on the germination of stratified F. americana embryos? If Amo-1618 did not inhibit germination, does this mean that stratification did not result in an increased level of endogenous GA in the embryos? Explain carefully.

DORMANCY OF SEEDS OF WHITE ASH (FRAXINUS AMERICANA)

REPORT

Name _____ Section _____ Date _____

6. Describe the effect of ABA on the germination of stratified F. americana embryos. Relate your result to what has been discovered by others about the natural occurrence of ABA in Fraxinus seeds, the comparative amounts of ABA in F. ornus and dormant F. americana seeds, and the effect of stratification on ABA level in F. americana seeds.

7. What conclusions can reasonably be made from this experiment regarding the roles of specific growth substances in the regulation of dormancy in F. americana seeds?

8. In what ways does seed dormancy (and the chilling requirement to over-come dormancy) and the absence of such a dormancy mechanism contribute to the adaptation of F. americana and F. ornus, respectively, to their natural habitats? Briefly describe the natural geographic distribution and the general climatic conditions for each of the two species.

EXERCISE 12

CHEMICAL BREAKING AND INDUCTION OF BUD DORMANCY

Introduction

Temporary suspension of growth, or deceleration to barely perceptible rates, called <u>dormancy</u>, is a common feature of the ontogeny of seed plants. Dormancy is a phenomenon of profound biological significance, because it provides a means by which plants are enabled to survive periods of environmental conditions which would be adverse or lethal to plants in an active state of growth. Thus, in the case of annual plants, dormancy of seeds is commonplace. In biennials and perennials, dormancy of the buds of established plants and of their seeds and storage organs (e.g. bulbs, corms and tubers) are typical events in the life cycles.

Unfortunately, considerable confusion exists concerning the terminology relating to dormancy phenomena. But, by generally acceptable connotation, <u>dormancy</u> in the broad sense refers to <u>any temporary suspension of active growth</u>. Dormancy which is <u>imposed by unfavorable physical environmental</u> conditions such as low temperature or moisture stress, which often is immediately reversible upon the plant's again experiencing favorable conditions, is called <u>quiescence</u>, <u>imposed</u> dormancy, <u>temporary</u> dormancy or, sometimes, in the case of perennials, <u>summer</u> dormancy. <u>True</u> dormancy, <u>rest</u> or <u>innate</u> dormancy refer to <u>a state of temporarily suspended growth which is due to internal conditions</u>. In the case of true dormancy or rest, growth is suspended even under physical environmental conditions which apparently are favorable for growth. It is true dormancy or rest, hereafter called simply dormancy, of buds of woody plants with which this experiment is concerned.

Most, indeed probably all, woody plants form dormant buds at some stage in the annual cycle of growth. In temperate zone species, to which further discussion is largely restricted, dormant buds typically are formed in late summer or autumn, and the whole plant enters a dormant or resting phase. Tropical woody plants also form dormant buds, but not all the branches of a given shoot may enter dormancy at the same time, and dormancy in these plants often is difficult or impossible to correlate with specific environmental conditions. Rather, dormancy in such plants may display a periodicity which is under endogenous control.

The most important physical environmental factor in the induction of bud dormancy in temperate zone woody plants is daylength, or, more specifically, the relative duration and timing of daily light and dark periods. In the majority of species which have been investigated (angiosperms and gymnosperms), long days promote vegetative growth, and short days (and long nights) bring about the cessation of growth and the onset of bud dormancy. The leaves are the organs which directly respond to the photoinductive stimulus, and phytochrome is the photoreceptor pigment.

Not to be ignored, however, is the fact that the onset of bud dormancy appears not to be simply explained as a photoinduction phenomenon in some species of woody plants. A number of kinds of fruit trees (e.g. <u>Pyrus</u>, <u>Malus</u> and <u>Prunus</u>) and some others (e.g. <u>Fraxinus</u> <u>excelsior</u> and certain members of the Oleaceae) show very little response to daylength. And in some species (e.g. larch) in which young trees exhibit conspicuous photoperiodic responses, older trees may not. In these plants the onset of dormancy is far from completely understood.

Emergence from dormancy in most woody plants is dependent upon exposure to low temperature ($\sim$ 1-10° C) for periods of approximately 250 to more than 1,000 hours, depending on the species. Some species (e.g. Betula pubescens and Acer pseudoplatanus) can be induced to break dormancy by transfer to long-day conditions, but this is not true in the case of the majority of species. Although chilling is necessary to break dormancy in most species, warm ambient air temperatures are necessary for the actual resumption of growth. Frequently, the chilling requirement is met by midwinter, but the buds fail to resume growth until warmer spring temperatures prevail.

Just as it is important to understand the environmental stimuli which evoke the onset and breaking of bud dormancy, it is of great interest to understand the changes in internal conditions which are causally related to the annual cycle of growth and dormancy. Especially prominent among these changes are alterations in the balance of hormones, particularly the growth-stimulating gibberellins (GAs) and the growth-inhibiting hormone abscisic acid (ABA) (Fig. 12-1).

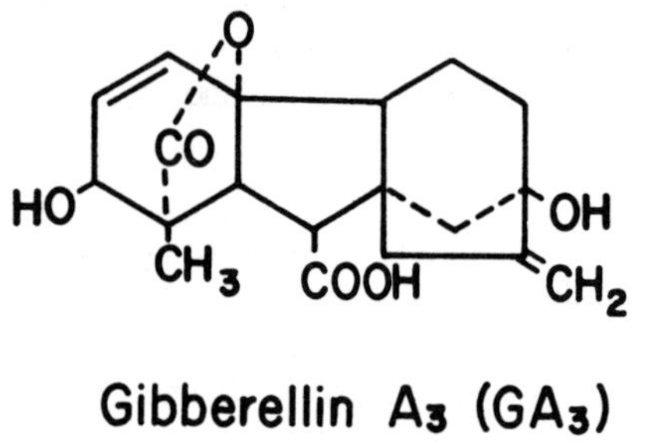

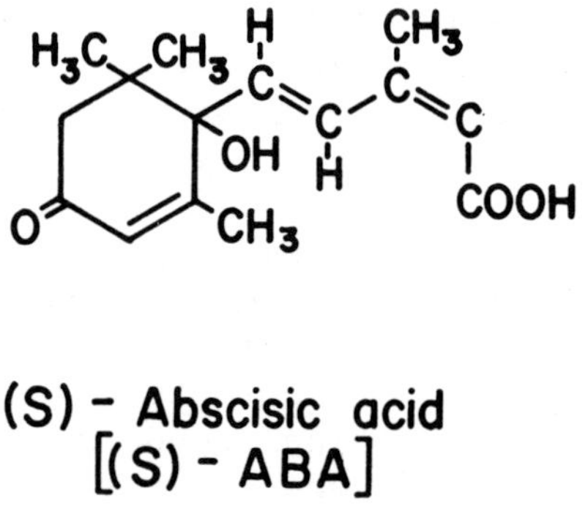

Gibberellin A₃ (GA₃)   (S) - Abscisic acid [(S) - ABA]

Fig. 12-1. Structures of gibberellin A_3 and (S)-abscisic acid.

A prevalent current concept is that the annual cycle of bud growth and dormancy in temperate zone woody plants actually is regulated by a balance between endogenous growth inhibitors and growth-stimulating hormones, with ABA and GA being of particular, if not exclusive, importance in at least some species. Thus it appears that the induction of bud dormancy by short days (and long nights) is due to accumulation of ABA (but see reference 14). Release from dormancy, stimulated by low temperature and/or long days, seems to be determined primarily by an increase in endogenous GA and secondarily by a decrease in endogenous ABA.

The purpose of this experiment is to investigate the effects of exogenous GA and ABA on breaking and induction, respectively, of dormancy in a woody plant under opposing photoperiodic conditions.

Materials and Methods

Approximately 30 to 40 uniform, young filbert (Corylus avellana L.)[1] trees (or other suitable species) with shoots 30 to 50 cm long will be utilized. The plants will be growing in individual pots filled with soil and will be kept in a greenhouse. The manner in which the experiment is started will depend upon the condition (dormant or actively growing) of the plants.

If the plants are dormant at the time the experiment is to be started, as expected, they should be exposed to photoperiodic conditions conducive to keeping the plants dormant; that is, relatively short (8- to 10-hour) days and long (14- to 16-hour) nights. If the natural photoperiodic conditions are not adequate, they should be controlled. Treatment of the plants should be performed as follows. Using a razor blade or sharp scalpel, cut off the apical 2-3 mm of the terminal bud of each plant, so as to provide a direct route of entry of treatment solution to the growing point. Then, with a micrometer buret, apply 10 µl of solution containing GA_3 (1 mg/ml), 5% ethanol and 0.05% Tween 20 to each of half the plants. Treat each of the remaining plants with 10 µl of solution containing no GA_3. Observe the plants at each laboratory period for the next 3 to 4 weeks. Repeat the chemical treatment one week after the first treatment if there is no evidence of resumption of active growth by the plants treated with GA. Record at each time of observation the number of plants in each group which have obviously broken dormancy. Plot a graph of the percentage of plants in each group which break dormancy versus time after treatment with GA.

At the conclusion of the above experiment, expose all the actively growing plants from the group treated with GA to photoperiodic cycles consisting of 16- to 20-hour photoperiods and 4- to 8-hour dark periods, which should be favorable for the continuation of active growth. Divide the plants equally into two groups. Measure the length of the newly formed portion of the primary shoot of each plant in both groups at this time and again one week later. Record the initial mean growth rate (cm/week) for each group. At the end of the one week under long-day conditions, spray the plants in one group with a solution containing 2 g/liter ABA, 25% ethanol and 0.1% Tween 20. Spray the other group of plants with solution lacking ABA. Observe the plants at each laboratory period, and measure

---

[1]Seeds of filbert (Corylus avellana) and numerous other tree species can be purchased from F. W. Schumacher Company, Sandwich, Massachusetts 02563, and Herbst Brothers Seedsmen, Incorporated, Brewster, New York 10509.

the growth rate (cm increase in length/week) of the newly formed part of the main shoot of each plant in both groups for the succeeding 3 to 4 weeks. Plot a graph showing the mean growth rate (cm/week) of the plants in each group versus time after treatment with ABA. At the end of the experiment, record the percentage of plants in each group which are exhibiting conspicuous senescence of the leaves and the percentage which have set terminal buds.

If the plants are actively growing at the time the experiment is to be started, the two parts of the experiment should, of course, be done in opposite order to the manner they are described above.

References

1. Addicott, F. T. and J. L. Lyon. 1969. Physiology of abscisic acid and related substances. Ann. Rev. Plant Physiol. 20: 139-164.

2. Addicott, F. T., J. L. Lyon, K. Ohkuma, W. E. Thiessen, H. R. Carns, O. E. Smith, J. W. Cornforth, B. V. Milborrow, G. Ryback and P. F. Wareing. 1968. Abscisic acid: a new name for abscisin II (dormin). Science 159: 1493.

3. Aspinall, D., L. G. Paleg, and F. T. Addicott. 1967. Abscisin II and some hormone-regulated plant responses. Australian J. Biol. Sci. 20: 869-882.

4. Chrispeels, M. J. and J. E. Varner. 1967. Hormonal control of enzyme synthesis: on the mode of action of gibberellic acid and abscisin in aleurone layers of barley. Plant Physiol. 42: 1008-1016.

5. Cornforth, J. W., B. V. Milborrow, G. Ryback, and P. F. Wareing. 1965. Identity of sycamore 'dormin' with abscisin II. Nature 205: 1269-1270.

6. Eagles, C. F. and P. F. Wareing. 1963. Experimental induction of dormancy in Betula pubescens. Nature 199: 874-875.

7. Eagles, C. F. and P. F. Wareing. 1964. The role of growth substances in the regulation of bud dormancy. Physiol. Plantarum 17: 697-709.

8. El-Antably, H. M. M., P. F. Wareing, and J. Hillman. 1967. Some physiological responses to d,l abscisin (dormin). Planta 73: 74-90.

9. Galston, A. W. and P. J. Davies. 1970. Control Mechanisms in Plant Development. Prentice-Hall, Englewood Cliffs, New Jersey.

10. Greathouse, D. C., W. M. Laetsch, and B. O. Phinney. 1971. The shoot-growth rhythm of a tropical tree, _Theobroma cacao_. Am. J. Botany 58: 281-286.

11. Hemberg, T. 1949. Growth inhibiting substances in buds of _Fraxinus_. Physiol. Plantarum 2: 37-44.

12. Kawase, M. 1961. Growth substances related to dormancy in _Betula_. Proc. Am. Soc. Hort. Sci. 78: 532-544.

13. Kramer, P. J. and T. T. Kozlowski. 1960. Physiology of Trees. McGraw-Hill Book Company, New York.

14. Lenton, J. R., V. M. Perry, and P. F. Saunders. 1972. Endogenous abscisic acid in relation to photoperiodically induced bud dormancy. Planta 106: 13-22.

15. Leopold, A. C. and P. E. Kriedemann. 1975. Plant Growth and Development. 2nd Ed. McGraw-Hill Book Company, New York.

16. Moore, T. C. 1979. Biochemistry and Physiology of Plant Hormones. Springer-Verlag, New York.

17. Pearson, J. A. and P. F. Wareing. 1969. Effect of abscisic acid on activity of chromatin. Nature 221: 672-673.

18. Perry, T. O. 1971. Dormancy of trees in winter. Science 171: 29-36.

19. Phillips, I. D. J. 1971. Introduction to the Biochemistry and Physiology of Plant Growth Hormones. McGraw-Hill Book Company, New York.

20. Phillips, I. D. J. and P. F. Wareing. 1958. Effect of photoperiodic conditions on the level of growth inhibitors in _Acer pseudoplatanus_. Naturwissenschaften 13: 317.

21. Ohkuma, K., J. L. Lyon, F. T. Addicott, and O. E. Smith. 1963. Abscisin II, an abscission-accelerating substance from young cotton fruit. Science 142: 1592-1593.

22. Robinson, P. M., P. F. Wareing, and T. H. Thomas. 1963. Isolation of the inhibitor varying with photoperiod in _Acer pseudoplatanus_. Nature 199: 875-876.

23. Salisbury, F. B. and C. W. Ross. 1978. Plant Physiology. 2nd Ed. Wadsworth Publishing Company, Belmont, California.

24. Scih, C. Y. and L. Rappaport. 1970. Regulation of bud rest in tubers of potato, _Solanum tuberosum_ L. VII. Effect of abscisic acid and gibberellic acids on nucleic acid synthesis in excised buds. Plant Physiol. 45: 33-36.

25. Sloger, C. and B. E. Caldwell. 1970. Response of cultivars of soybean to synthetic abscisic acid. Plant Physiol. 46: 634-635.

26. Thomas, T. H., P. F. Wareing, and P. M. Robinson. 1965. Action of sycamore 'dormin' as a gibberellin antagonist. Nature 205: 1270-1272.

27. Van Overbeek, J. 1966. Plant hormones and regulators. Science 152: 721-731.

28. Van Overbeek, J., J. E. Loeffler, and M. I. R. Mason. 1967. Dormin (abscisin II), inhibitor of plant DNA synthesis? Science 156: 1497-1499.

29. Vegis, A. 1964. Dormancy in higher plants. Ann. Rev. Plant Physiol. 15: 185-224.

30. Walton, D. C., G. S. Soofi, and E. Sondheimer. 1970. The effects of abscisic acid on growth and nucleic acid synthesis in excised embryonic bean axes. Plant Physiol. 45: 37-40.

31. Wareing, P. F. 1954. Growth studies in woody species. IV. The locus of photoperiodic perception in relation to dormancy. Physiol. Plantarum 7: 261-277.

32. Wareing, P. F. 1956. Photoperiodism in woody plants. Ann. Rev. Plant Physiol. 7: 191-214.

33. Wareing, P. F. and I. D. J. Phillips. 1970. The Control of Growth and Differentiation in Plants. Pergamon Press, New York.

34. Wareing, P. F. and D. L. Roberts. 1956. Photoperiodic control of cambial activity in Robinia pseudacacia L. New Phytol. 55: 356-366.

Special Materials and Equipment Required
Per Laboratory Section

(30 to 50) Young Filbert (Hazelnut) (Corylus avellana) trees (or other deciduous tree species native to temperate region of North America), preferably dormant, with shoots approximately 30 to 50 cm long (Seeds available from F. W. Schumacher Company, Sandwich, Massachusetts 02563)

(Approximately 250 ml) ABA solution (containing 2 mg ABA per ml, 25% ethanol and 0.1% Tween 20) (RS-ABA available from Hoffman-LaRoche and Company, 1599 Factor Avenue, San Leandro, California 94577)

(Approximately 50 ml) GA solution (containing 1 mg GA_3 per ml, 5% ethanol and 0.05% Tween 20)

(1) Micrometer buret (e.g. 0.2-ml capacity micrometer buret available from Roger Gilmont Instruments, Incorporated, Vineland, New Jersey)

(1) Controlled-environment growth chamber or greenhouse with temperature and photoperiod controls

Recommendations for Scheduling

It is recommended that, if at all possible, this experiment be initi-
ated with dormant plants. No difficulty is likely to be encountered in
breaking bud dormancy with exogenous GA, whereas chemical induction of bud
dormancy with exogenous ABA may not be entirely successful, depending upon
the species used and other factors. Also, because of the extended period
required to perform both parts of the experiment, it may not be possible
to complete the entire experiment in a one-term course. The breaking of
bud dormancy with exogenous GA itself constitutes a very worthwhile experiment. In any case, the experiment should be started early in the term.
Because of the requirement of considerable space to grow the plants, it is
recommended that the experiment be conducted as a class project in each
laboratory section or whole class.

Duration: 7 to 10 weeks; approximately 1 hour required during
period(s) when treatments are administered; brief (30 minutes or less)
periodic observations during other periods.

EXERCISE 12

CHEMICAL BREAKING AND INDUCTION OF BUD DORMANCY

REPORT

Name _____ Section _____ Date _____

<u>Results of treatment with GA:</u>

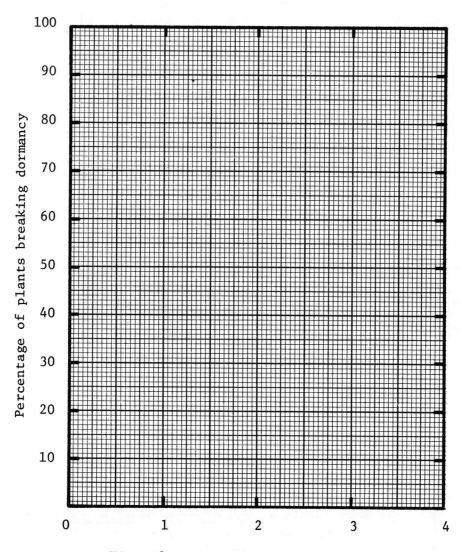

Time after treatment with GA (weeks)

CHEMICAL BREAKING AND INDUCTION OF BUD DORMANCY

REPORT

<u>Results of treatment with ABA:</u>

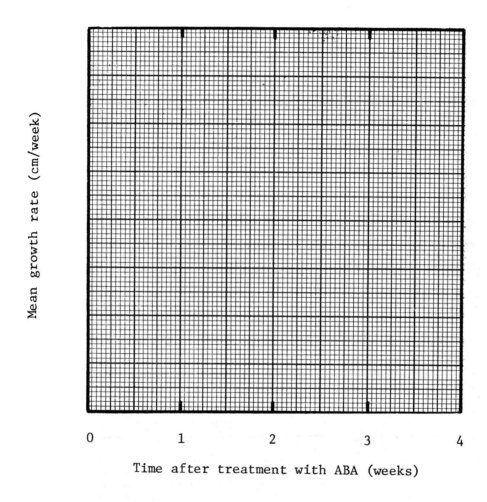

Time after treatment with ABA (weeks)

Plant group	% of plants at end of experiment	
	Exhibiting leaf senescence	Setting terminal bud
Controls	_____	_____
ABA	_____	_____

CHEMICAL BREAKING AND INDUCTION OF BUD DORMANCY

REPORT

Name _____ Section _____ Date _____

Questions

1. Why were the plants treated with GA kept under short photoperiods and those treated with ABA under long photoperiods?

2. State possible reasons why the particular methods of application of GA and ABA (direct application to terminal bud and spraying of whole plant, respectively) were chosen.

3. Discuss the similarities and differences between the responses of the plants to GA and ABA and the natural behavior of the plants as they emerge from dormancy and go dormant under natural conditions.

CHEMICAL BREAKING AND INDUCTION OF BUD DORMANCY

REPORT

Name _____ Section _____ Date _____

4. Based on the results of the ABA treatment in this experiment, does it
    seem more than a mere coincidence that ABA was originally discovered
    as a substance which both induces bud dormancy and accelerates senes-
    cence and abscission of leaves? Explain.

5. Identify some other naturally-occurring growth inhibitors besides ABA
    which may be causally involved in the dormancy of specific kinds of
    buds and seeds.

EFFECTS OF ABSCISIC ACID AND BENZYLADENINE ON
GROWTH AND DORMANCY IN LEMNA

Introduction

Growth and development of higher plants is regulated by several
kinds of hormones, including most prominently the growth-stimulating
auxins, gibberellins, and cytokinins and the growth-inhibiting hormone
abscisic acid. Ethylene, unique among the naturally-occurring growth
substances in being a gas under physiological conditions, also is considered generally to be a hormone which influences numerous physiological
processes. Although chemical characterization or identification of such
substances has not yet been made, it appears highly probable that there
also is one or more specific flowering hormones among species of angiosperms.

Developmental events--such as seed germination, onset and breaking
of bud dormancy and floral initiation--and rates of growth are governed
largely by delicate balances among the endogenous hormones. For example,
if the concentration of growth-inhibiting substances such as abscisic
acid becomes sufficiently great, relative to the concentrations of specific growth-stimulating hormones, a plant may temporarily cease to grow
and become dormant. Although many growth-inhibiting substances have been
identified in plants, abscisic acid (ABA) (Fig. 13-1) is the only inhibitor which currently is generally regarded as a hormone.

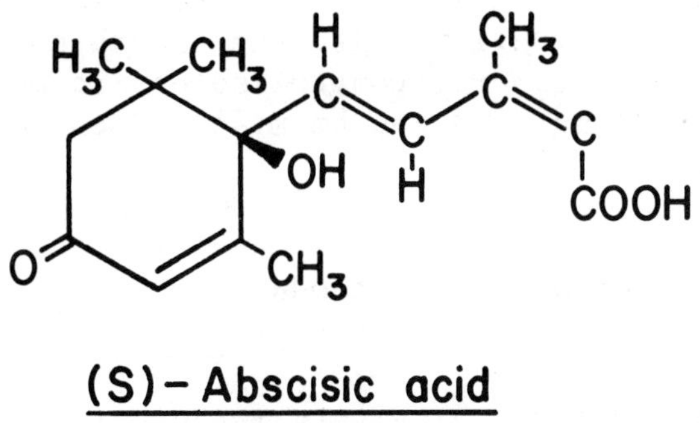

(S) - Abscisic acid

Fig. 13-1. Structure of (S)-abscisic
acid.

Experiments conducted in recent years, involving applications of
growth-regulating chemicals to whole plants and isolated plant parts,
have revealed numerous interesting interactions between growth-stimulating
hormones and ABA. Thus ABA is antagonistic (non-competitively) to certain
actions of gibberellins, such as induction of synthesis of particular enzymes in aleurone cells of barley grains, stimulation of elongation (in
the presence of auxin) of excised stem and coleoptile sections, and promotion of germination of particular kinds of seeds. In some of these

cases the inhibitory action of ABA can be counteracted by gibberellin and, in others, by cytokinin.

The purpose of this experiment is to investigate the inhibitory effect of ABA on the growth of the small floating aquatic angiosperm _Lemna minor_ and the counter-effect of a particular synthetic cytokinin, benzyladenine (BA) (Fig. 13-2). _Lemna_ is a nearly ideal choice of plant

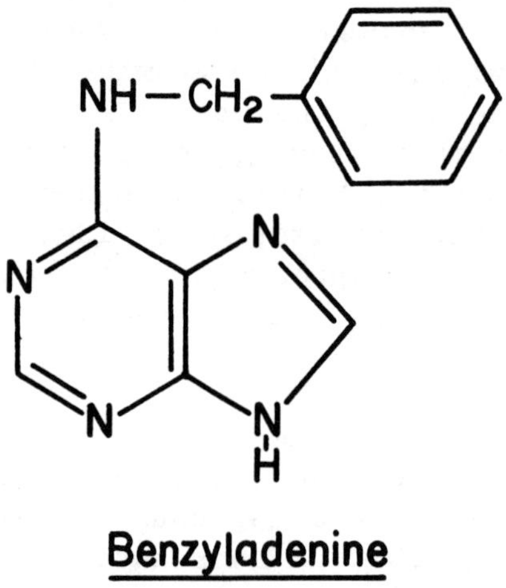

Benzyladenine

Fig. 13-2. Structure of a synthetic
cytokinin, benzyladenine.

for this kind of experiment because it is of very widespread natural distribution, hence large quantities frequently can be collected conveniently. Because of its small size and aquatic habit, the plants can be grown in relatively little space under uniform, controlled environmental conditions. Growth is vegetative, by budding, and can be measured conveniently as increase in fresh weight.

Materials and Methods[1]

Each group of students prepare 9 culture vessels (e.g. Carolina culture dishes, 3.5 inch O.D.) each containing 50 ml of one of the following solutions of growth regulator(s) dissolved in complete mineral nutrient solution (see Exercise 22 - "Mineral Nutrition of Sunflowers" for composition):

---

[1]Adapted from: Van Overbeek, J., J. E. Loeffler, and M. I. R. Mason. 1967. Dormin (abscisin II), inhibitor of plant DNA synthesis? Science 156: 1497-1499.

Solution	BA conc. (mg/liter)	ABA conc. (mg/liter)
1	0.0	0.0
2	0.0	0.1
3	0.0	1.0
4	0.1	0.0
5	0.1	0.1
6	0.1	1.0
7	1.0	0.0
8	1.0	0.1
9	1.0	1.0

The culture solutions may be conveniently prepared by adding 0.1 or 1.0 ml of a stock solution of ABA or BA at a concentration of 50 mg/liter to 50 ml of nutrient solution. Mark the level of solution on each vessel with a wax pencil or felt pen, so that the volume can be restored to initial level with the appropriate solution during each succeeding laboratory period. Label each culture vessel with information identifying the group of students and the treatment.

Using a forceps or glass rod, remove 180 Lemna minor plants from the large culture provided. Wash the sample of plants three times with complete nutrient solution, blot very gently on paper toweling, and weigh the plants on an analytical balance. Next, randomly place 20 plants in each of the culture vessels. Record the average initial fresh weight of the 20-plant samples.

Place all the culture vessels in a growth chamber under constant fluorescent light (∿ 2,500 ft-c) and constant temperature (22±1° C), or alternatively, under fluorescent lamps on the laboratory bench, or near a window in the laboratory. Observe the plants at each laboratory period; restore the volume of each solution to the original level with the appropriate solution. Seven to 10 days after the start of the experiment, make final, careful observations of the plants in all groups. Remove each sample of plants, blot gently on paper toweling and weigh to the nearest tenth milligram. Record the results in a table.

References

1. Addicott, F. T. and J. L. Lyon. 1969. Physiology of abscisic acid and related substances. Ann. Rev. Plant Physiol. 20: 139-164.

2. Addicott, F. T., J. L. Lyon, K. Ohkuma, W. E. Thiessen, H. R. Carns, O. E. Smith, J. W. Cornforth, B. V. Milborrow, G. Ryback, and P. F. Wareing. 1968. Abscisic acid: a new name for abscisin II (dormin). Science 159: 1493.

3. Aspinall, D., L. G. Paleg, and F. T. Addicott. 1967. Abscisin II and some hormone-regulated plant responses. Australian J. Biol. Sci. 20: 869-882.

4. Chrispeels, M. J. and J. E. Varner. 1966. Inhibition of gibber-
    ellic acid induced formation of α-amylase by abscisin II.
    Nature 212: 1066-1067.

5. Chrispeels, M. J. and J. E. Varner. 1967. Hormonal control of
    enzyme synthesis: on the mode of action of gibberellic acid
    and abscisin in aleurone layers of barley. Plant Physiol.
    42: 1008-1016.

6. Cornforth, J. W., B. V. Milborrow, G. Ryback, and P. F. Wareing.
    1965. Identity of sycamore 'dormin' with abscisin II.
    Nature 205: 1269-1270.

7. El-Antably, H. M. M., P. F. Wareing, and J. Hillman. 1967. Some
    physiological responses to d,1 abscisin (dormin). Planta 73:
    74-90.

8. Galston, A. W. and P. J. Davies. 1969. Hormonal regulation in
    higher plants. Science 163: 1288-1297.

9. Galston, A. W. and P. J. Davies. 1970. Control Mechanisms in Plant
    Development. Prentice-Hall, Englewood Cliffs, New Jersey.

10. Khan, A. A., L. Andersen, and T. Gaspar. 1970. Abscisic acid-
    induced changes in nucleotide composition of rapidly labeled
    ribonucleic acid species of lentil root. Plant Physiol. 46:
    494-495.

11. Leopold, A. C. and P. E. Kriedemann. 1975. Plant Growth and
    Development. 2nd Ed. McGraw-Hill Book Company, New York.

12. Letham, D. S. 1969. Cytokinins and their relation to other
    phytohormones. BioScience 19: 309-316.

13. Moore, T. C. 1979. Biochemistry and Physiology of Plant Hormones.
    Springer-Verlag, New York.

14. Pearson, J. A. and P. F. Wareing. 1969. Effect of abscisic acid
    on activity of chromatin. Nature 221: 672-673.

15. Phillips, I. D. J. 1971. Introduction to the Biochemistry and
    Physiology of Plant Growth Hormones. McGraw-Hill Book
    Company, New York.

16. Salisbury, F. B. and C. W. Ross. 1978. Plant Physiology. 2nd Ed.
    Wadsworth Publishing Company, Belmont, California.

17. Thomas, T. H., P. F. Wareing, and P. M. Robinson. 1965. Action
    of sycamore 'dormin' as a gibberellin antagonist. Nature
    205: 1270-1272.

18. Van Overbeek, J. 1966. Plant hormones and regulators. Science
    152: 721-731.

19. Van Overbeek, J., J. E. Loeffler, and M. I. R. Mason. 1967.
    Dormin (abscisin II), inhibitor of plant DNA synthesis?
    Science 156: 1497-1499.

20. Wareing, P. F. and I. D. J. Phillips. 1970. The Control of
    Growth and Differentiation in Plants. Pergamon Press,
    New York.

Special Materials and Equipment Required
Per Team of 3-8 Students

(Approximately 180) Duckweed (Lemna minor) plants, preferably taken
    directly from a pure culture. Cultures of Lemna minor can be pur-
    chased from Carolina Biological Supply Company, Burlington, North
    Carolina 27215.
(9) Culture vessels (e.g. Carolina culture vessels, 3.5 inch O.D.,
    Carolina Biological Supply Company)
(450 ml) Complete nutrient solution
(Approximately 10 ml) Stock solution of BA at concentration of 50 mg/liter
(Approximately 10 ml) Stock solution of RS-ABA at concentration of 50
    mg/liter (RS-ABA available from Hoffman-LaRoche and Company, 1599
    Factor Avenue, San Leandro, California 94577)
(1) Controlled-environment facility, nearly constant conditions of 22 to
    26° C and light intensity of approximately 2,500 ft-c

Recommendations for Scheduling

It is recommended that students in each laboratory section work in
three to six teams, each team performing the whole experiment. To expedite setting up the experiment, it is recommended that the nutrient
solution and stock solutions of BA and ABA be prepared in advance of the
period when the experiment is to be started.

Duration: 7 to 10 days; approximately 1 1/2 hours required during
first period and less than 30 minutes each successive period until period
when experiment is terminated, when approximately 1 hour is required.

EFFECTS OF ABSCISIC ACID AND BENZYLADENINE ON GROWTH AND DORMANCY IN LEMNA

REPORT

Name _____ Section _____ Date _____

Results — final fresh weights of Lemna cultures:[1]

BA conc. (mg/liter)	ABA conc. (mg/liter)		
	0.0	0.1	1.0
0.0	_____ mg	_____ mg	_____ mg
0.1	_____	_____	_____
1.0	_____	_____	_____

Observations on coloration, morphology and frequency of budding:

---

[1]Average initial fresh weight of cultures = _____ mg.

EXERCISE 13

EFFECTS OF ABSCISIC ACID AND BENZYLADENINE ON GROWTH AND DORMANCY IN LEMNA

REPORT

Name _____ Section _____ Date _____

Questions

1. Describe the effects of BA and ABA independently on the growth and general appearance and morphology of Lemna.

2. Were any of the plants treated with ABA truly "dormant"? Explain.

3. Did BA at either or both of the concentrations tested completely counteract the effect of the highest concentration of ABA? Explain.

EXERCISE 13

EFFECTS OF ABSCISIC ACID AND BENZYLADENINE ON GROWTH AND DORMANCY IN LEMNA

REPORT

Name _____ Section _____ Date _____

4. Whereas you measured growth as increments in fresh weight, can you suggest other parameters which could have been used in this experiment?

5. What would you predict would have happened to the Lemna plants treated with ABA had they merely been transferred to fresh nutrient solution free of ABA? Would the plants perhaps have resumed a normal growth rate? Explain.

6. Do you suppose auxins and gibberellins, at particular concentrations, might be equally as effective as BA in counteracting the growth-inhibitory action of ABA on Lemna? Explain.

ROLES OF AUXINS AND CYTOKININS IN APICAL DOMINANCE

Introduction

The term <u>apical dominance</u> refers to the inhibition of growth of sub-tending lateral (axillary) buds by a growing shoot apex. Some degree of apical dominance is found in all seed plants, and the phenomenon obvious-ly is of profound significance in determining growth form. In some herbaceous plants [e.g. pea (<u>Pisum sativum</u>) and sunflower (<u>Helianthus annuus</u>)] apical dominance is quite strong, the lateral buds commonly are suppressed, and the shoot thus develops as a single main axis or monopodium. In other species [e.g. tomato (<u>Lycopersicum esculentum</u>) and potato (<u>Solanum tuberosum</u>)] apical dominance is much weaker and the shoot is extensively branched. Sometimes apical dominance is only partial, in which case some of the lateral buds develop while others remain fully inhibited.

In certain species of vascular plants--e.g. <u>Ginkgo biloba</u>, the Maidenhair tree, and <u>Cercidiphyllum japonicum</u>, the Katsura tree of Japan--there is a special and very interesting case of lateral buds which open but which do not elongate appreciably. Instead they produce "short shoots" as opposed to the more usual "long shoots." When a vigorous terminal bud develops first it establishes dominance over the laterals below. This inhibition is transmitted by the auxin produced in the elongating terminal ("long") shoot. In these plants a lateral "short shoot" behaves funda-mentally like an inhibited bud of an herbaceous plant, differing in that it opens while an inhibited lateral bud does not. Therefore the occurr-ence of "short shoots" and "long shoots" on a plant such as <u>Ginkgo</u> is regarded as a special instance of the phenomenon of apical dominance.

Apical dominance clearly is very important in determining the growth habits of woody perennials. However, some confusion has existed in the past because of attempts to explain the adult forms of trees and shrubs by too strict and sometimes erroneous analogy with apical dominance in herbaceous plants. The relationship between bud inhibition and form in woody plants is much more complex than bud inhibition in herbaceous plants because of the time sequence in the formation and release of lateral buds. The life cycles of woody perennials are characterized by periods of dor-mancy. A dormant apical bud does not exert an inhibitory influence on the subtending lateral buds, and more than one bud may develop when dormancy is broken and growth resumes. This may partially explain why so much of the branching of woody perennials occurs at the beginning of a period of active growth. Apical dominance most certainly is directly involved in determining the pattern of growth of individual twigs and branches. It would be well to restrict the term to the pattern of bud inhibition in <u>currently elongating twigs</u>. To denote the complex of physiological conditions giving rise to excurrent and decurrent growth forms of whole trees and shrubs, the term <u>apical control</u> has been proposed.

While the mechanism of apical dominance is incompletely understood at present, it is known that the phenomenon involves an antagonism between

153

two kinds of hormones, namely <u>auxins</u> and <u>cytokinins</u>. Thus, auxin, commonly secreted by the growing apex, is inhibitory to the release of buds from
apical dominance, and this action of auxin is antagonized by cytokinin,
which is probably synthesized in the suppressed lateral bud itself. Lateral buds typically are relieved of inhibition in a growing shoot by
removal of the shoot apex. Inhibition of the lateral buds is reinstated
if auxin is applied to the cut surface of the decapitated stem. Local
application of cytokinin directly to inhibited lateral buds causes a
release from inhibition, However, buds thus released from apical dominance do not elongate as much as uninhibited buds released by decapitation
of the shoot apex, unless they are treated with auxin following release
with cytokinin. Thus, in the intact plant, the inhibited lateral bud may--
as paradoxical as it might at first appear--fail to grow because of a
deficiency of <u>both</u> cytokinin <u>and</u> auxin.

Correlated directly with the effects of auxin and cytokinin on
lateral bud inhibition and release therefrom are well-documented effects
of these two kinds of hormones on differentiation of vascular tissues
between axillary buds and the primary stele of the main shoot axis.
Histological studies have revealed that inhibited lateral buds of some
plants lack well-developed vascular connections with the vascular system
of the main stem. This observation suggests, of course, that inhibited
buds receive a deficient supply of metabolites, which evidence shows
actually to be the case. Removal of the apical bud from a growing shoot
leads to rapid development of vascular connections between the released
bud and the vascular system of the main shoot axis. As would be predicted,
auxin applied to a decapitated shoot tip inhibits the formation of vascular connections, and cytokinin, applied locally to the lateral buds,
promotes the differentiation of the vascular traces.

The purpose of this experiment is to investigate the roles of auxin
(indoleacetic acid, IAA) (Fig. 14-1) and cytokinin (kinetin) (Fig. 14-1)
in the control of lateral bud development in an herbaceous dicot which
normally exhibits strong apical dominance.

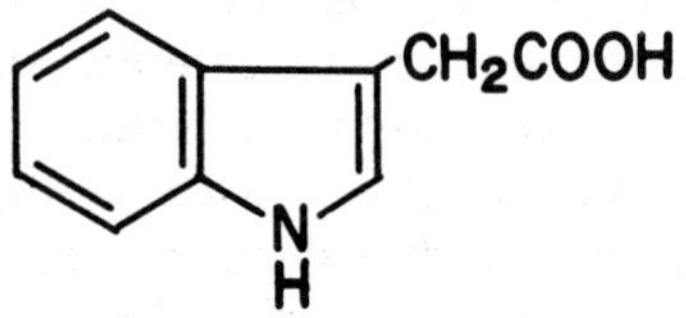

Indoleacetic acid
(IAA)

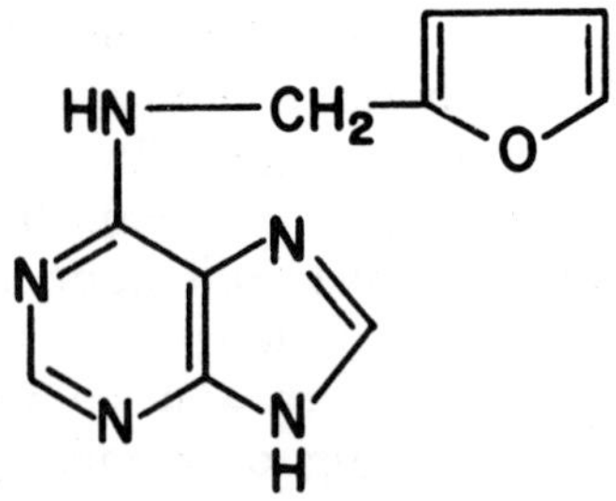

Kinetin

Fig. 14-1. Structures of indoleacetic acid, an
auxin, and kinetin, a cytokinin.

Materials and Methods[1]

Each team of students will require 8 groups (or 6 groups; see footnote 5) of approximately 10 Alaska pea seedlings. Alaska pea (<u>Pisum sativum</u>) seeds should be surface-disinfected by first washing with vigorous agitation in a solution of 0.5% NaOCl (10% commercial Clorox) and then rinsing several times with distilled water. Plant 15 to 20 seeds each in 6-inch plastic pots filled with vermiculite and saturate the vermiculite with distilled water. Culture the plants in a growth chamber, if available, or elsewhere, under continuous light at an intensity of approximately 800 ft-c and a constant temperature of approximately 20° C. Irrigate the seedlings as necessary during the first week of culture with distilled water and thereafter with a complete mineral nutrient solution (see Exercise 22 - "Mineral Nutrition of Sunflowers" for composition).

On the 6th or 7th day after planting, thin the plants to leave approximately 10 uniform plants in each group. Treat one group in each of the following ways:

1. Leave plants intact; <u>carefully</u> apply (with a micrometer buret) 6 µl of kinetin solution[2] <u>to the apex</u> of the lateral bud at the second node (second or uppermost scale leaf node) of each plant.[3]

2. Same as "1", except apply 6 µl of 50% ethanol + 8% carbowax 1500 only (no kinetin) to the lateral bud.

3. Decapitate (with a razor blade) each seedling at the top of the third internode; apply plain lanolin to the cut surface; and apply 6 µl of kinetin solution to the apex of the lateral bud at the second node.

---

[1] Adopted from: Sachs, T. and K. V. Thimann. 1964. Release of lateral buds from apical dominance. Nature 201: 939-940 <u>and</u> Sachs, T. and K. V. Thimann. 1967. The role of auxins and cytokinins in the release of buds from dominance. Am. J. Botany 54: 136-144.

[2] The kinetin solution should contain 330 mg/1 kinetin, 50% ethanol and 8% carbowax 1500. The mixture must be heated in boiling water for about 10 minutes to dissolve the kinetin.

[3] It is of utmost importance in all cases where treatment of the lateral bud is prescribed to apply the solution very carefully to the apex of the bud. For best quantitative results, it is recommended that all the lateral buds except the treated one be carefully removed with a dissecting needle or other device from plants of all groups, although this latter operation definitely is not essential to a successful experiment.

4. Same as "3", except apply 6 μl of 50% ethanol + 8% carbowax only (no kinetin) to the lateral bud.

5. Decapitate each seedling (as in "3"); apply lanolin containing IAA[4] to the cut surface of the stem; and apply 6 μl of kinetin solution to the apex of the lateral bud at the second node.

6. Same as "5", except apply 6 μl of 50% ethanol + 8% carbowax 1500 only (no kinetin) to the lateral bud.

Optional parts of experiment:[5]

7. Leave plants intact; apply 6 μl of kinetin solution to the lateral bud at the second node; and apply 6 μl of 1000 mg/1 IAA in 50% ethanol to the apex of the lateral bud three days following treatment with kinetin.

8. Same as "7", except apply 6 μl of 50% ethanol + 8% carbowax only (no kinetin) to the lateral bud, followed 3 days later by 6 μl of 1000 mg/1 IAA in 50% ethanol.

Seven days after the first treatment measure the length of the bud (or branch) at the second node of every plant in each group; record the number of plants in each group in which the lateral bud at the second node is released from inhibition; remove the buds (or branches) from the second node, pool all those for each group, and weigh. Record the data in a table.

References

1. Ali, A. and R. A. Fletcher. 1970. Xylem differentiation in inhibited cotyledonary buds of soybeans. Can. J. Botany 48: 1139-1140.

---

[4]Dissolve 100 mg of crystalline indole-3-acetic acid in 1 to 2 ml of 95% or absolute ethanol, and mix thoroughly with 100 g of anhydrous lanolin in a petri dish. Heat the mixture on a hot plate to expedite mixing.

[5]Treatments 7 and 8 should be viewed as optional parts of the experiment, which some instructors may choose to delete. The predicted result-- auxin-induced growth of the cytokinin-released bud (cf. treatments 1 and 7)--is not always readily obtained. In some varieties the auxin effect on bud growth apparently is very sensitive to the stage of development of the bud at the time of treatment; in all cases success is absolutely dependent on application of auxin directly and exclusively to the apex of the bud. Treatments 1 through 6 should give clear and unequivocal results, and if attempted without success, the results of treatments 7 and 8 should not be regarded as detracting from an otherwise successful experiment.

2. Brown, C. L., R. G. McAlpine, and P. P. Kormanik. 1967. Apical dominance and form in woody plants: a reappraisal. Am. J. Botany 54: 153-162.

3. Galston, A. W. and P. J. Davies. 1970. Control Mechanisms in Plant Development. Prentice-Hall, Englewood Cliffs, New Jersey.

4. Gunckel, J. E. and K. V. Thimann. 1949. Studies of development in long shoots and short shoots of Ginkgo biloba L. III. Auxin production in shoot growth. Am. J. Botany 36: 145-151.

5. Gunckel, J. E., K. V. Thimann, and R. H. Wetmore. 1949. Studies of development in long shoots and short shoots of Ginkgo biloba L. IV. Growth habit, shoot expression and the mechanism of its control. Am. J. Botany 36: 309-316.

6. Moore, T. C. 1979. Biochemistry and Physiology of Plant Hormones. Springer-Verlag, New York.

7. Phillips, I. D. J. 1969. Apical dominance. Pp. 165-202 in: M. B. Wilkins, Ed. The Physiology of Plant Growth and Development. McGraw-Hill Publishing Company, New York.

8. Phillips, I. D. J. 1971. Introduction to the Biochemistry and Physiology of Plant Growth Hormones. McGraw-Hill Book Company, New York.

9. Sachs, T. and K. V. Thimann. 1964. Release of lateral buds from apical dominance. Nature 201: 939-940.

10. Sachs, T. and K. V. Thimann. 1967. The role of auxins and cytokinins in the release of buds from dominance. Am. J. Botany 54: 136-144.

11. Salisbury, F. B. and C. W. Ross. 1978. Plant Physiology. 2nd Ed. Wadsworth Publishing Company, Belmont, California.

12. Sorokin, H. P., and K. V. Thimann. 1964. The histological basis for inhibition of axillary buds in Pisum sativum and the effects of auxins and kinetin on xylem development. Protoplasma 59: 326-350.

13. Steward, F. C. and A. D. Krikorian. 1971. Plants, Chemicals and Growth. Academic Press, New York.

14. Thimann, K. V., T. Sachs, and K. N. Mathur. 1971. The mechanism of apical dominance in Coleus. Physiol. Plantarum 24: 68-72.

15. Titman, P. W. and R. H. Wetmore. 1955. The growth of long and short shoots in Cercidiphyllum. Am. J. Botany 42: 364-372.

16. Wickson, M. E. and K. V. Thimann. 1958. The antagonism of auxin and kinetin in apical dominance. Physiol. Plantarum 11: 62-74.

17. Zimmermann, M. H. and C. L. Brown. 1971. Trees, Structure and Function. Springer-Verlag, New York. (Pp. 17-29).

Special Materials and Equipment Required
Per Team of 3-8 Students

(Approximately 150) Alaska pea (Pisum sativum) seeds (Commercial seed companies or local store)
(8) Plastic pots (approximately 6 inch diameter at top) and drain saucers
(Approximately 8 liters) Vermiculite, sand or other planting medium
(Stock supply) Complete nutrient solution
(1 or 2) Micrometer buret (e.g. 0.2-ml capacity micrometer buret available from Roger Gilmont Instruments, Incorporated, Vineland, New Jersey 08360)
(Approximately 5 ml) Kinetin solution (containing 330 mg/liter kinetin, 50% ethanol and 8% carbowax 1500)
(Approximately 5 ml) 50% ethanol-8% carbowax 1500 solution
(Approximately 5 ml) IAA liquid solution (containing 1 g IAA per liter of 50% ethanol)
(Approximately 5 ml) 50% ethanol
(Approximately 5 g) IAA in lanolin (containing 100 mg IAA per 100 g anhydrous lanolin)
(Approximately 5 g) Plain anhydrous lanolin
(1) Controlled-environment facility

Recommendations for Scheduling

It is suggested that students work in teams of 3 to 8 members each, with each team performing the entire experiment. Alternatively, each team can be assigned to do only a designated part of the experiment. Considerable time can be saved in setting up the experiment if all solutions are prepared in advance.

Duration: 13 or 14 days; approximately 1 hour required for planting of seeds during first period, approximately 1 hour required during period when treatments are made, approximately 1 hour required during period when experiment is terminated, and less than 15 minutes per period required at other times.

EXERCISE 14

ROLES OF AUXINS AND CYTOKININS IN APICAL DOMINANCE

REPORT

Name _____ Section _____ Date _____

Treatment	No. plants	No. plants on which bud[1] released	Average length of bud[1] (mm)	Average fresh weight per bud[1] (mg)
1. Intact; kinetin on lateral bud				
2. Intact; no kinetin on lateral bud				
3. Decapitated; lanolin on cut surface; kinetin on lateral bud				
4. Decapitated; lanolin on cut surface; no kinetin on lateral bud				
5. Decapitated; lanolin-IAA on cut surface; kinetin on lateral bud				
6. Decapitated; lanolin-IAA on cut surface; no kinetin on lateral bud				
7. Intact; kinetin on lateral bud; IAA 3 days later on bud				
8. Intact; no kinetin on lateral bud; IAA on bud as in "7"				

[1]Bud at second scale leaf node.

159

EXERCISE 14

ROLES OF AUXINS AND CYTOKININS IN APICAL DOMINANCE

REPORT

Name _____ Section _____ Date _____

Questions

1. Summarize concisely the effects of kinetin and of IAA which were observed in this experiment.

2. How closely does the release of buds from inhibition by application of exogenous cytokinin represent a natural process? Explain.

3. What is a possible explanation for the different effects of IAA on bud inhibition and release, which were observed with groups 5, 6, 7 and 8?

ROLES OF AUXINS AND CYTOKININS IN APICAL DOMINANCE

REPORT

Name _____ Section _____ Date _____

4. For what reasons were alcohol and carbowax added to the kinetin solution?

5. What effect do you suppose kinetin might have had on the plants in group 3 had the cytokinin been applied to the cut surface of the decapitated stem instead of to the apex of the lateral bud?

EXTRACTION AND BIOASSAY OF GIBBERELLINS
FROM FUSARIUM MONILIFORME

Introduction

The gibberellins (GAs) are one of the major groups of growth-promoting hormones which, in common with auxins, cytokinins, various growth inhibitors (e.g. abscisic acid) and ethylene, play essential roles in the regulation of growth and development of seed plants.

While GAs are now universally regarded as hormonal substances participating in the regulation of normal growth and development, the discovery of GAs occurred as a result of certain investigations of diseased or abnormal growth. Before the turn of the century Japanese farmers were concerned about a disease of rice called bakanae, which was responsible for considerable losses of this staple crop. The diseased plants exhibit excessive elongation of the stems and leaf sheaths and often fall and become lodged before reaching maturity.

Plant pathologists discovered that an ascomycete, Gibberella fujikuroi, specifically the asexual stage, Fusarium moniliforme, was responsible for the disease. In 1926, E. Kurosawa performed experiments which resulted in the discovery of the substances which later would be named gibberellins by T. Yabuta and T. Hayashi. Kurosawa's basic experiment was to grow G. fujikuroi in a liquid culture medium, and to subsequently apply samples of the culture filtrate to healthy rice seedlings. The filtrate evoked characteristic bakanae symptoms, thus demonstrating a chemical basis for the disease.

By the 1930's the Japanese had succeeded in crystallizing and identifying certain GAs from the fungus. However, scientists outside Japan did not become aware of this important work immediately as it was published because of the breakdown of communications during World War II years. Not until the the 1950's did research on GAs begin in England and the United States. The earliest of these studies concerned effects of GAs isolated from G. fujikuroi cultures on seed plants, especially angiosperms. Many dramatic and extremely interesting effects were noted, the most common being an often fantastic promotion of stem elongation in dwarf angiosperms (e.g. dwarf varieties of corn, peas and beans). GAs also were found to: (a) cause certain kinds of plants which normally have a long-day requirement or dual long-day and vernalization requirements for flowering to flower under non-inductive conditions; (b) break dormancy of many kinds of seeds and buds that normally have chilling requirements; and (c) have numerous other interesting effects.

Then, in 1956, it was demonstrated by C. A. West and B. O. Phinney at the University of California at Los Angeles, and independently about the same time by Margaret Radley in England, that GA-like substances are in fact natural constituents of certain higher plants. Now it is firmly established that GAs are natural hormones of seed plants generally. Indeed, GAs are now known to occur in some representatives of angiosperms, gymnosperms, ferns, algae and fungi. They may be ubiquitous in the plant kingdom, although their significance in lower forms is far from completely understood.

By 1978, 52 GAs had been identified and their structural formulas determined. Generally, analyses indicate two to several GAs in each type of plant investigated, with variations in both kinds and amounts of GAs being common in different parts of a single plant.

Chemically the GAs are 19- or 20-carbon tetracyclic diterpenoid compounds having a basic gibberellane skeleton (Fig. 15-1). Most, but not all, of the GAs also have a lactone ring attached to the A ring of the gibberellane skeleton. Other variations are in the number of carboxyl groups, the presence or absence of a double bond in the A ring, the number and position of hydroxyl groups, and the nature of the substituent on the D ring.

The purpose of this experiment is to determine the amount of gibberellin produced by a weighed sample of the fungus in a specified period by: (1) preparing an extract of the fungus culture filtrate and testing the growth-promoting activity of the extract on dwarf pea seedlings; and (2) comparing the response to the extract with a standard dose-response curve obtained by treating dwarf peas with a range of known dosages of gibberellin A$_3$.

Materials and Methods

A. Planting of seeds. Twelve to 14 days before the extract is to be prepared plant 10 plastic pots filled with vermiculite with seeds of a dwarf variety of garden peas in the greenhouse or a growth chamber programmed to provide a 16-hour photoperiod at approximately 20° C and 1,000 ft-c and an 8-hour dark period at 16° C. Also plant one pot of seeds of a tall pea variety. Irrigate the plants as often as needed with a complete mineral nutrient solution (see Exercise 22 - "Mineral Nutrition of Sunflowers" for composition).

B. Culturing the fungus. Approximately 6 days prior to the date the extract is to be prepared, transfer a portion (∿ 1 g fresh weight) of Fusarium moniliforme mycelium from a potato agar culture[1] to a 500-ml

---

[1]Potato dextrose agar culture medium is prepared according to a procedure reported by Stodola et al. (Arch. Biochem. Biophys. 54: 240-245. 1955) as follows: Peel and dice 200 g of potato tubers. Boil the potato tissue for 5 minutes in 500 ml of distilled water, filter through cheesecloth, and bring the filtrate back to the original volume with distilled water. To the filtrate add 100 ml of distilled water containing 20 g dextrose, 0.2 g CaCO$_3$ and 0.2 g MgSO$_4$; then add 400 ml of hot water in which 15 g agar has been melted. Pour 125 ml or 200 ml of the hot medium into each 250- or 500-ml Erlenmeyer flask, plug the flasks with cotton, and autoclave at 15 psi for 20 minutes. (When maintained at room temperature, one flask contains enough mycelium for one GA extraction procedure within 4 to 6 weeks after inoculation.)

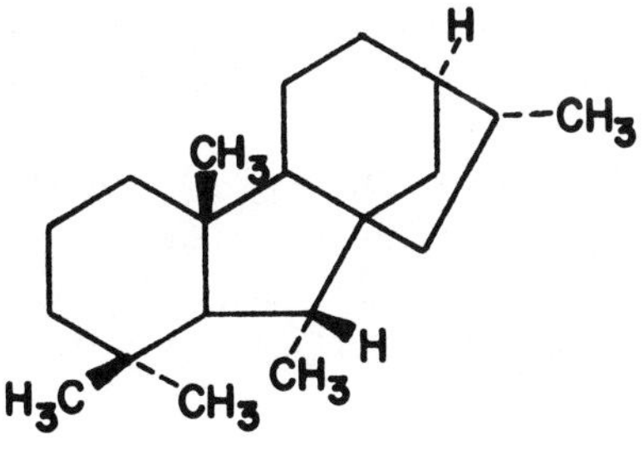

Gibberellane

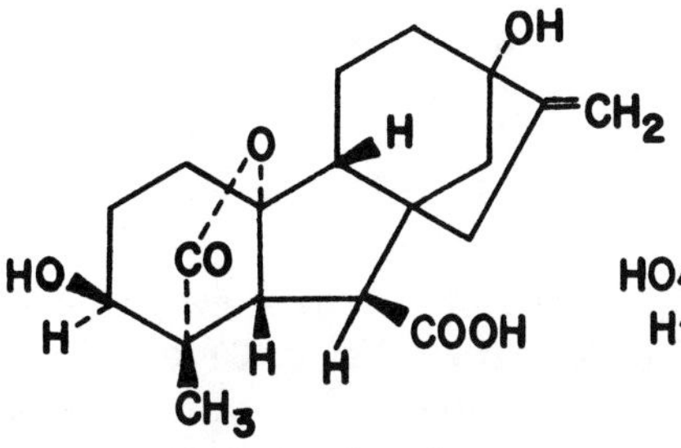

Gibberellin A₁ Gibberellin A₃ Gibberellin A₅

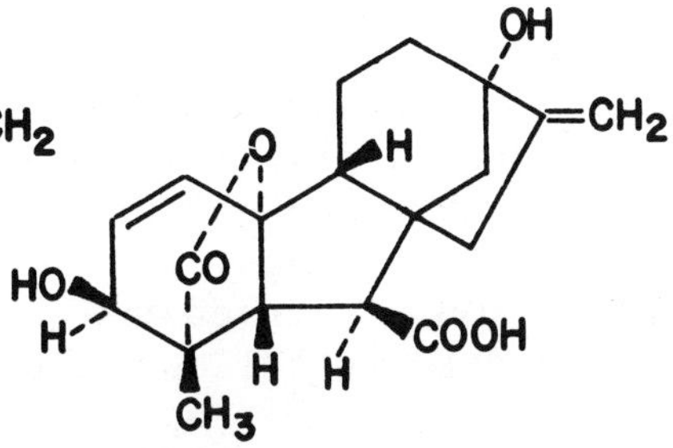

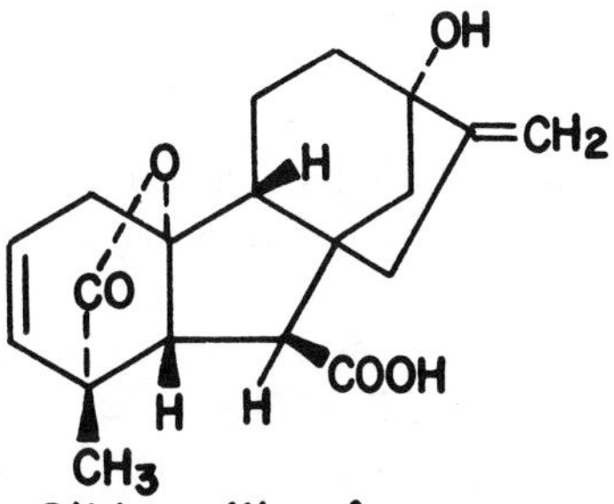

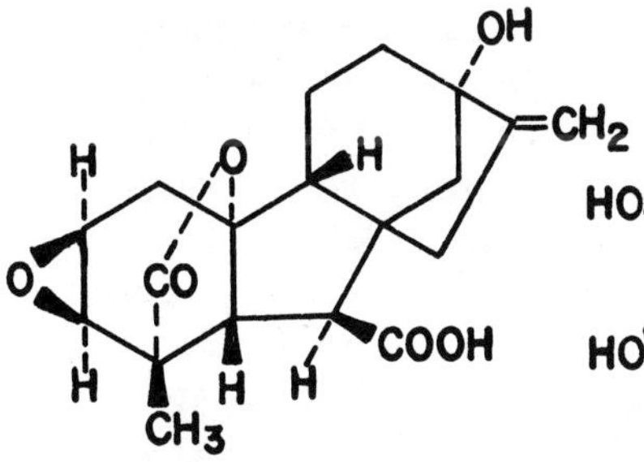

Gibberellin A₆ Gibberellin A₈ Gibberellin A₁₀

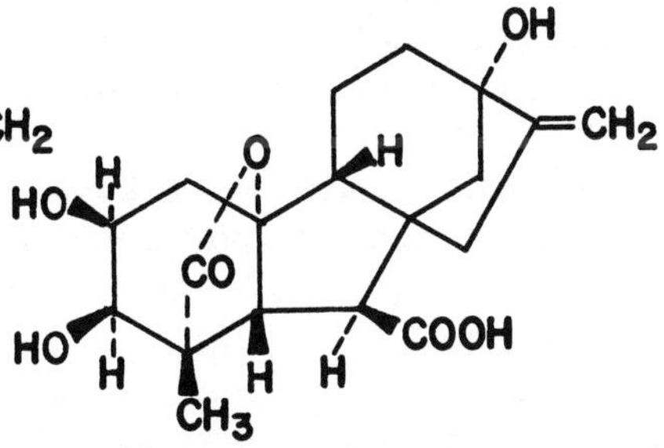

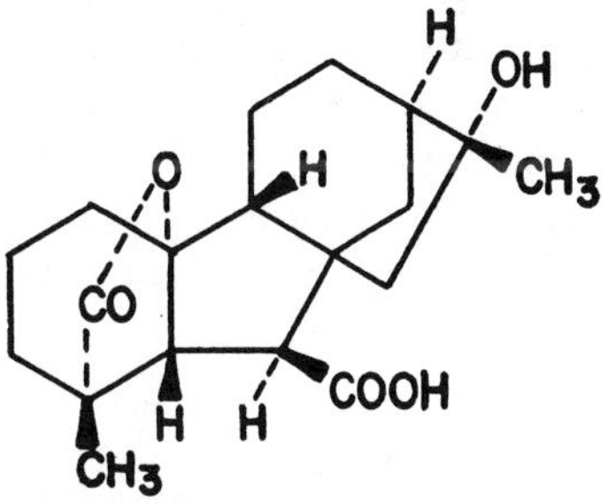

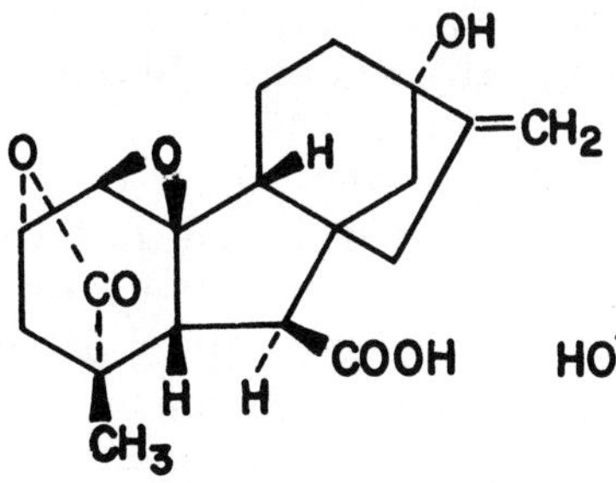

Gibberellin A₁₁ Gibberellin A₁₃ Gibberellin A₁₅

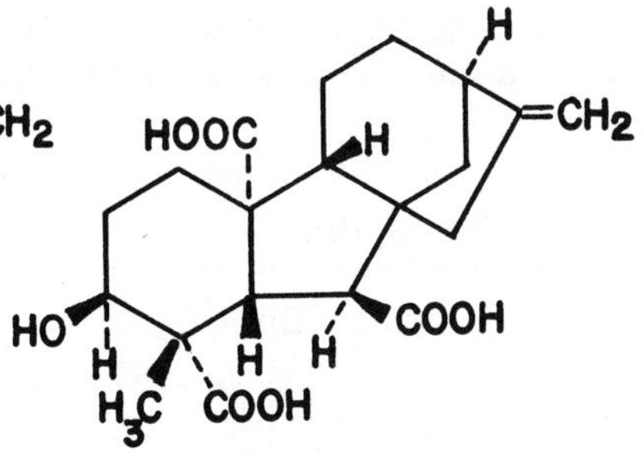

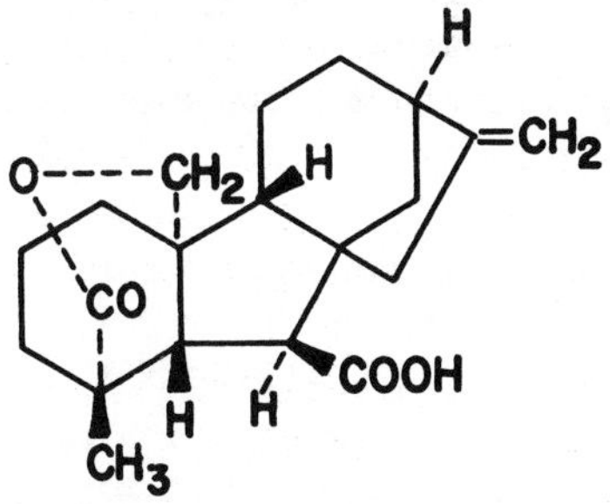

Fig. 15-1. Structures of the gibberellane skeleton and of nine gibberellins. While gibberellane is the currently recognized parent skeleton, the GAs actually are sterically analogous to the enantiomer of gibberellane, ent-gibberellane. (Redrawn from Lang, 1970.)

165

Erlenmeyer flask containing 100-ml of liquid nutrient solution.[2] Normally, the entire mycelium from one 500-ml culture flask should be homogenized aseptically in 50 ml of sterile liquid culture medium, and 8 ml of the inoculum should be aseptically transferred to each flask containing liquid culture medium. Care must be taken to avoid contamination of the fungus cultures with other microorganisms. Plug the flask with cotton and attach the flask to a mechanical shaker. Allow the culture to grow for approximately 6 days.

C. Extraction procedure. Weigh to the nearest tenth milligram a disk of filter paper to fit a selected Büchner funnel. Filter the fungus culture, using a Büchner funnel and a filtration flask attached to an aspirator. Save the filtrate. Remove the filter paper containing the mycelium, place in a petri dish and dry in an oven at 80° C. Record the dry weight of the mycelium. Check the pH of the filtrate; if the pH is not within the range pH 2.5 to 3.0, adjust it to within that range by carefully adding dilute HCl or KOH. Extract the aqueous filtrate twice with equal volumes of ethyl acetate as follows: determine the volume of filtrate; then pour into a 250- or 500-ml separatory funnel. Add an equal volume of ethyl acetate, shake vigorously and permit the water and ethyl acetate phases to separate. Drain off the lower water fraction and save. Then drain out the ethyl acetate phase and save. Repeat the extraction of the water fraction with ethyl acetate. Combine the two ethyl acetate fractions and place them in a 500-ml round-bottom flask. Attach the flask to a rotary type evaporator, immerse bottom of flask in a water bath at 30 to 40° C and evaporate the extract to dryness. Take up the residue in 10-ml of 25% ethanol and 0.05% Tween 20. Prepare approximately 10 ml of GA_3 in 25% ethanol containing 0.05% Tween 20, at the following concentrations: 0.01, 0.1, 1.0, 10.0, and 100 µg/10 µl. Also prepare a solution containing 25% ethanol and 0.05% Tween 20 only.

D. Bioassay procedure. Thin the dwarf pea seedlings to leave about 10 or more uniform seedlings per pot. Measure the shoot heights of the plants from the cotyledonary node to the highest visible node. Using a micrometer buret, apply 10 µl, containing 0.01, 0.1, 1.0, 10.0, or 100 µg of GA_3, to the shoot apex of each plant in 5 groups. Apply 10 µl of extract to each plant in a sixth group, 10 µl of extract diluted 1:9 with 0.05% Tween 20 solution to each plant in a seventh group, and 10 µl of extract diluted 1:99 to each plant in an eighth group. Also apply 10 µl of 0.05% Tween 20 solution alone to each plant in a control group. Five days to one week later, again measure the shoot heights of the plants. Calculate the mean increment in shoot length for each group. Plot a standard curve by plotting, on semi-log paper, increase in shoot length in cm versus dosage of GA_3. Compare the responses of the three groups of extract-treated plants, and interpolate to determine the mean concentration of gibberellin (as GA_3 equivalents) in the extract; express as x µg/µl. Then calculate the total gibberellin produced by a measured amount of dry weight of mycelium during the specified period in shaken liquid culture.

_____

[2]The liquid culture medium consists of 2.0% glucose, 0.3% $MgSO_4 \cdot 7H_2O$, 0.3% NH_4Cl and 0.3% KH_2PO_4 in distilled water.

References

1. Barendse, G. W. M., H. Kende, and A. Lang. 1968. Fate of radio-active gibberellin A_1 in maturing and germinating seeds of peas and Japanese morning glory. Plant Physiol. 43: 815-822.

2. Chin, T. Y. and J. A. Lockhart. 1965. Translocation of applied gibberellin in bean seedlings. Am. J. Botany 52: 828-833.

3. Galston, A. W. and P. J. Davies. 1969. Hormonal regulation in higher plants. Science 163: 1288-1297.

4. Galston, A. W. and P. J. Davies. 1970. Control Mechanisms in Plant Development, Prentice-Hall, Englewood Cliffs, New Jersey.

5. Hedden, P., J. MacMillan, and B. O. Phinney. 1978. The metabolism of the gibberellins. Ann. Rev. Plant Physiol. 29: 149-192.

6. Jones, R. L. 1968. Aqueous extraction of gibberellins from pea. Planta 81: 97-105.

7. Jones, R. L. and A. Lang. 1968. Extractable and diffusible gibber-ellins from light- and dark-grown pea seedlings. Plant Physiol. 43: 629-634.

8. Jones, R. L. and I. D. J. Phillips. 1964. Agar-diffusion tech-nique for estimating gibberellin production by plant organs. Nature 204: 497-499.

9. Jones, R. L. and I. D. J. Phillips. 1966. Organs of gibberellin synthesis in light-grown sunflower plants. Plant Physiol. 41: 1381-1386.

10. Kato, J. 1958. Nonpolar transport of gibberellin through pea stem and a method for its determination. Science 128: 1008-1009.

11. Kende, H. 1967. Preparation of radioactive gibberellin A_1 and its metabolism in dwarf peas. Plant Physiol. 42: 1612-1618.

12. Lang, A. 1970. Gibberellins: structure and metabolism. Ann. Rev. Plant Physiol. 21: 537-570.

13. Lockhart, J. A. 1957. Studies on the organ of production of the natural gibberellin factor in higher plants. Plant Physiol. 32: 204-207.

14. McComb, A. J. 1961. "Bound" gibberellin in mature runner bean seeds. Nature 192: 575-576.

15. McComb, A. J. 1964. The stability and movement of gibberellic acid in pea seedlings. Ann. Botany 28: 669-687.

16. Moore, T. C. 1967. Gibberellin relationships in the 'Alaska' pea (*Pisum sativum*). Am. J. Botany 54: 262-269.

17. Moore, T. C. 1979. Biochemistry and Physiology of Plant Hormones. Springer-Verlag, New York.

18. Musgrave, A. and H. Kende. 1970. Radioactive gibberellin A5 and its metabolism in dwarf peas. Plant Physiol. 45: 56-61.

19. Paleg, L. G. 1965. Physiological effects of gibberellins. Ann. Rev. Plant Physiol. 16: 291-322.

20. Phillips, I. D. J. 1971. Introduction to the Biochemistry and Physiology of Plant Growth Hormones. McGraw-Hill Book Company, New York.

21. Phillips, I. D. J. and R. L. Jones. 1964. Gibberellin-like activity in bleeding sap of root systems of *Helianthus annuus* detected by a new dwarf pea epicotyl assay and other methods. Planta 63: 269-278.

22. Radley, M. 1958. The distribution of substances similar to gibberellic acid in higher plants. Ann. Botany 22: 297-307.

23. Radley, M. 1967. Site of production of gibberellin-like substances in germinating barley embryos. Planta 75: 164-171.

24. Salisbury, F. B. and C. W. Ross. 1978. Plant Physiology. 2nd Ed. Wadsworth Publishing Company, Belmont, California.

25. Sitton, D., A. Richmond, and Y. Vaadia. 1967. On the synthesis of gibberellins in roots. Phytochemistry 6: 1101-1105.

26. Skene, K. G. M. 1967. Gibberellin-like substances in root exudate of *Vitis vinifera*. Planta 74: 250-262.

27. Steward, F. C. and A. D. Krikorian. 1971. Plants, Chemicals and Growth. Academic Press, New York.

28. Stowe, B. B. and T. Yamaki. 1957. The history and physiological action of the gibberellins. Ann. Rev. Plant Physiol. 8: 181-216.

29. Wareing, P. F. and I. D. J. Phillips. 1970. The Control of Growth and Differentiation in Plants. Pergamon Press, New York.

30. Zweig, G., S. Yamaguchi, and G. W. Mason. 1961. Translocation of ^{14}C-gibberellin in red kidney bean, normal corn and dwarf corn. Adv. in Chem. Ser. 28: 122-134.

Special Materials and Equipment Required
Per Laboratory Section

(1) Flask culture of Gibberella fujikuroi (Fusarium moniliforme). An
    inoculum can be purchased from the American Type Culture Collection,
    12301 Parklawn Drive, Rockville, Maryland 20852; it is recommended
    that the strain designated "NRRL 2284 (Fusarium moniliforme)" be
    specified.
(Approximately 250) Seeds of a dwarf variety of pea (Pisum sativum), e.g.
    Carter's Daisy = Dwarf Telephone,[3] Progress No. 9, or Little Marvel
    (e.g. W. Atlee Burpee Company, Philadelphia, Pennsylvania, Clinton,
    Iowa or Riverside, California)
(Approximately 25) Seeds of a tall variety of pea (Pisum sativum), e.g.
    Alderman = Tall Dark-podded Telephone or Alaska[3] (W. Atlee Burpee
    Company)
(Stock supply) Complete mineral nutrient solution
(1) Controlled-environment facility for pea plants
(10 to 12) 6-inch plastic pots and saucers
(10 to 12 liters) Vermiculite
(1) Mechanical shaker
(1) Rotary-film evaporator and 500-ml round bottom flask
(1) Water bath
(1) Filtration flask, Büchner funnel, aspirator set-up
(1 to several) Micrometer buret (e.g. 0.2 ml capacity micrometer buret
    available from Roger Gilmont Instruments, Incorporated, Vineland,
    New Jersey 08360)
(1) Separatory funnel (250- or 500-ml)
(200 ml) Ethyl acetate
(Approximately 10 ml) Each of the following solutions of GA_3 in 25%
    ethanol and 0.05% Tween 20: 0, 0.01, 0.1, 1.0, 10.0 and 100 mg
    GA_3/10 ml

Recommendations for Scheduling

It is recommended that the experiment be conducted as a class project
in each laboratory section. The students should perform all the operations
described in Materials and Methods, except preparation of liquid culture
medium for the fungus and preparation of the GA solutions, which should
be done in advance of the periods in which these materials are needed.
If the experiment is adopted for repeated use in the course, the fungus
should be routinely propagated on potato dextrose agar medium (see Materials and Methods).

——————————————

[3]It is recommended that the two varieties Dwarf Telephone (= Carter's
Daisy) and Tall (Dark-podded) Telephone (= Alderman) be used, since these
two varieties differ only in the genes controlling shoot length. Alternatively, Progress No. 9 (dwarf) and Alaska (tall or normal) make a good
pair of varieties.

<u>Duration</u>: 17 to 21 days (providing agar cultures of <u>Gibberella</u> <u>fujikuroi</u> are available for sub-culturing in liquid nutrient medium); 30 minutes to 1 hour required at first laboratory period to plant pea seeds, 30 to 45 minutes required for sub-culturing the fungus, 1.5 to 2 hours required for GA extraction and treatment of pea plants, and 1 hour required during period experiment is terminated.

EXERCISE 15

EXTRACTION AND BIOASSAY OF GIBBERELLINS FROM FUSARIUM MONILIFORME

REPORT

Name _____ Section _____ Date _____

Results of dwarf pea bioassay:

Treatment (µg GA$_3$/plant)	No. plants	Mean measurements of shoot length (cm)		
		Initial	Final	Increase
0.00	_____	_____	_____	_____
0.01	_____	_____	_____	_____
0.10	_____	_____	_____	_____
1.00	_____	_____	_____	_____
10.00	_____	_____	_____	_____
100.00	_____	_____	_____	_____
E 1:99	_____	_____	_____	_____
E 1:9	_____	_____	_____	_____
E	_____	_____	_____	_____

Calculation of concentration of GA$_3$
equivalents in extract:

Exercise 15

Extraction and Bioassay of Gibberellins from <u>Fusarium moniliforme</u>

REPORT

Name _____ Section _____ Date _____

<u>Standard GA₃ dosage-response curve:</u>

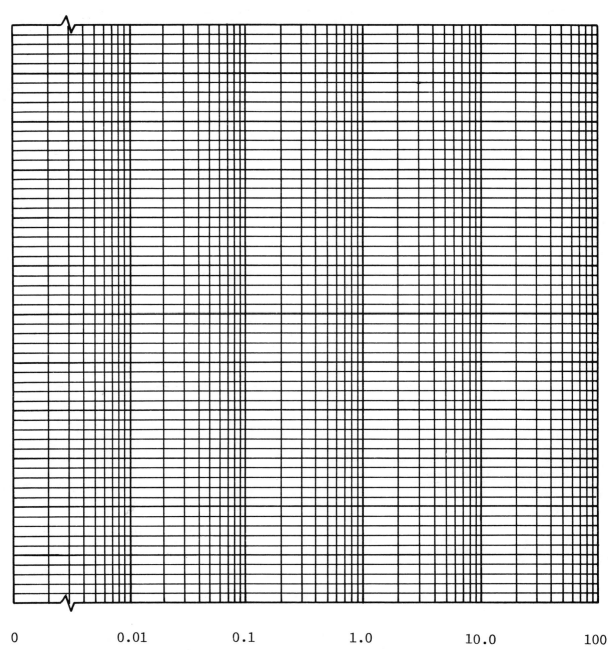

EXERCISE 15

EXTRACTION AND BIOASSAY OF GIBBERELLINS
FROM FUSARIUM MONILIFORME

REPORT

Name _____ Section _____ Date _____

Questions

1. What was the concentration (in $\mu g/\mu l$) of gibberellin (expressed as GA_3 equivalents) in the extract?

2. How much total gibberellin was produced by the shaken liquid culture in the specified growing period?

3. Is it likely that only one gibberellin was present in the extract? Explain.

4. What kinds of organisms are known to produce gibberellins?

5. Is it technically correct to speak of gibberellins as "hormones" of angiosperms? Explain.

EXERCISE 15

EXTRACTION AND BIOASSAY OF GIBBERELLINS
FROM FUSARIUM MONILIFORME

REPORT

Name _____ Section _____ Date _____

6. Is dwarfism in peas explainable on a gibberellin basis? Discuss.

EXERCISE 16

INDUCTION OF α-AMYLASE SYNTHESIS IN ALEURONE CELLS OF BARLEY GRAINS BY GIBBERELLIN

Introduction

Each kind of plant hormone is known to be capable of evoking a number of often quite diverse physiological effects. An intriguing question about any hormone is, "What is the biochemical mechanism of action of the hormone?" That is, what is the direct and primary effect of the hormone on metabolism, which is the basis for the often more readily observable physiological effects? Progress toward elucidating the biochemical mechanisms of action of all kinds of plant hormones--auxins, gibberellins, cytokinins, abscisic acid and ethylene--has been rapid in recent years. This experiment is designed to acquaint you with the mechanism of action of one kind of hormone, specifically the gibberellins.

In recent years, since 1960, it has been discovered that gibberellins (GAs) control the synthesis of particular enzymes. The most thoroughly investigated case is the GA_3-controlled synthesis of α-amylase (a starch hydrolyzing enzyme) in barley endosperm. It is well known that germinating seeds produce hydrolytic enzymes that digest the lipid, carbohydrate, and protein reserves of the storage tissues. Beginning with the discovery by L. G. Paleg and H. Yomo independently in 1960 that GA_3 increases the activity of α-amylase in barley endosperm, exciting progress has been made toward understanding the interactions of the various tissues in germinating barley grains and the role of GA in controlling the formation of enzymes.

In the intact, germinating barley grain the embryo produces GA, and this GA diffuses into the endosperm and surrounding aleurone layers where it evokes a large increase in α-amylase activity and the activities of some other enzymes as well, including β-1,3-glucanase, protease and ribonuclease. The increase in α-amylase activity is prevented by anaerobiosis, dinitrophenol, protein synthesis inhibitors and RNA synthesis inhibitors. It has been conclusively demonstrated that the increase in α-amylase activity in barley endosperm in response to added GA is due to de novo synthesis of the enzyme in the aleurone cells. All these data are consistent with the idea that GA controls the level of α-amylase and of other enzymes by controlling the synthesis of specific species of mRNA which in turn control the synthesis of specific enzymes.

Because the endosperm half of a barley grain (Fig. 16-1) produces several hydrolytic enzymes in response to added GAs, it is an attractive experimental system for a study of the mechanism of action of GA. The only living cells in the half grain (endosperm half dissected from end of grain containing embryo) are those of the aleurone layers. These are an apparently homogeneous collection of respiring, non-dividing cells which have the highly specialized function of producing and releasing some of those enzymes which are required to digest the starchy endosperm in preparation for its utilization by the growing embryo. The increase in

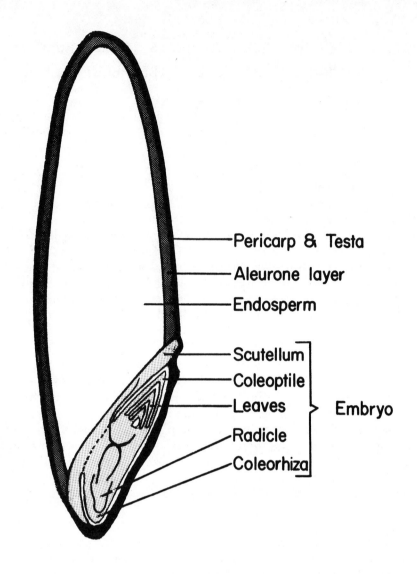

Fig. 16-1. Longitudinal section of a
huskless barley grain.

α-amylase activity in barley endosperm halves is absolutely dependent upon
added GA. Because of this, and the fact that the amount of α-amylase formed
is directly proportional to the amount of GA supplied over a wide range of
GA concentrations, a sensitive and highly specific bioassay for GA has been
developed which is based on the GA-induction of α-amylase.

Interestingly, as perhaps one would expect, nearly identical effects
of GA have been reported for other seeds and grains, including wild oat and
rice grains. Quite possibly, control of synthesis of particular enzymes
is the universal mode of action of GAs in plant growth regulation.

Materials and Methods[1]

Surface-disinfect dehusked (or naturally huskless)[2] barley grains by soaking with vigorous agitation in 5% NaOCl (sodium hypochlorite) in a stoppered flask for 3 hours at 25°. Then wash the grains at least 10 times with sterile distilled water. Next, incubate the disinfected grains in sterile distilled water for 20 to 24 hours at $\sim$ 3°. Either: (a) leave the grains in the flask, submerse with sterile water (leaving a considerable air-space above the grains) and place the flask on a mechanical shaker (if feasible); or (b) incubate the grains on sterile moist sand, or as a single layer of partially submersed grains, in a sterile covered container (e.g. petri dish). At the end of the period of imbibition, cut each grain transversely with a sterile scalpel or razor blade at a distance 4 mm from the distal end (opposite the embryo end). Then place two endosperm halves in each of a number of sterilized small glass vials (approximately 24 x 50 mm), each of which contains 1 ml of sterile water or GA_3 solution.[3] Prepare two vials for each of the following concentrations of GA_3: 0, 10^{-11}, 10^{-10}, 10^{-9}, 10^{-8}, and 10^{-7} g/ml. Also prepare in duplicate vials containing 10^{-8} g/ml GA_3 + 10 μg/ml cycloheximide and 10^{-8} g/ml GA_3 + 100 μg/ml actinomycin D. Thus, 8 test solutions and 16 vials are required. Close each vial with a sterile cork, set the vials in a test tube rack, mount the rack on a mechanical shaker (if convenient), and incubate all the vials for 24 to 48 hours at $\sim$ 30°.[4] (Note: In any assay such as this, contamination by microorganisms becomes a major concern. Every possible precaution must be taken to insure removal of microorganisms from the barley grains and to prevent subsequent introduction of microorganisms into the incubation mixtures.) After incubation the vials may be frozen (to stop the reaction) and stored in a freezer until such time as sugar measurements can conveniently be made, or the sugar assays can be made immediately.

On the day the incubate filtrates are to be analyzed for reducing

---

[1]Adopted from: Coombe, B. G., D. Cohen, and L. G. Paleg. 1967a. Barley endosperm bioassay for gibberellins. I. Parameters of the response system. Plant Physiol. 42: 105–112.

[2]Removal of husks can be done either by hand or by soaking the grains in 50% (by volume) H_2SO_4 for 4 hours at room temperature, followed by washing 15 to 20 times with autoclaved distilled water, with vigorous shaking to dislodge the husks. If dehusking is done with H_2SO_4, disinfection with 5% NaOCl can be omitted. Some investigators report better and more consistent results with husked cultivars than with naturally huskless cultivars.

[3]As a further precaution against interference by microbial contamination, chloramphenicol can be added to each test solution at a concentration of 25 μg/ml.

[4]One satisfactory alternative to the procedure described is to place disinfected embryo-less half-grains directly in the incubation solutions without prior soaking in sterile water. If this procedure is followed, the half-grains should be incubated about 48 hours before the incubate filtrates are analyzed. Another quite satisfactory alternative is the procedure described in detail in reference 18.

sugar, first prepare a standard curve for the colorimetric analysis using standard glucose solutions. Pipet duplicate 1-ml samples of glucose solutions at concentrations of 0, 10^{-7}, 10^{-6}, 10^{-5}, 3.3×10^{-5}, 10^{-4}, 3.3×10^{-4} and 10^{-3} g/ml into a series of 16 test tubes. Then add to each test tube 1.0 ml of <u>Somogyi's alkaline copper reagent</u>.[5] Place all 6 test tubes in a boiling water bath for precisely 10 minutes. Cool for 5 minutes in a cold water bath. Next, add to each test tube 1.0 ml of <u>arseno-molybdate reagent</u>[6] and mix the contents of each tube thoroughly. <u>CAUTION</u>: Be very careful pipeting the poisonous arseno-molybdate reagent; use a pro-pipet! Now dilute the contents of each test tube to a total volume of 10 ml, and read the absorbance at 560 nm in a colorimeter or spectrophoto-meter. Use the solution containing distilled water only (no glucose) as the <u>blank</u>. Standardize the colorimeter with this blank and use it for <u>all</u> the colorimetry. Record the absorbance values in a table. Plot, on log-log graph paper, a standard curve for the glucose solutions as "Absorbance" (on the ordinate) versus "Glucose concentration (g/ml)" (on the abscissa).

Next, prepare to analyze the incubate filtrates for reducing sugar. Add about 1 g of Amberlite IR120 (H^+) ion exchange resin and 9 ml of distilled water to each incubate, re-stopper the vial, and shake for 5 minutes. Filter each resin suspension through 1 thickness of Whatman No. 1 filter paper; this is most conveniently accomplished by inserting a cone of filter paper in the vial and pipeting from inside the cone. Then make <u>preliminary assays</u> for reducing sugars in the incubate filtrate from one of the vials containing 10^{-8} g/ml GA_3 by the following procedure. Pipet duplicate samples of 0.10, 0.25, 0.50 and 1.00 ml of filtrate into 8 test tubes. Make up the volume of each sample to 1.0 ml by adding the appropriate volume of distilled water to 6 of the test tubes. Then add to each test tube 1.0 ml of Somogyi's alkaline copper reagent. Place all 6 test tubes in a boiling water bath for precisely 10 minutes. Cool for 5 minutes in a cold water bath. Next, add to each test tube 1.0 ml of arseno-molybdate reagent and mix the contents of each tube thoroughly. Dilute the contents of each test tube to a total volume of 10 ml, and read the absorbance at 560 nm in a colorimeter or spectrophotometer. Determine which of the original volumes of incubate filtrate (0.10, 0.25, 0.50, or 1.00 ml) yields an absorbance value nearest 1.0. <u>Use this size</u>

---

[5]Somogyi's alkaline copper reagent: 24 g anhydrous Na_2CO_3
12 g NaK tartrate
16 g $NaHCO_3$
4 g $CuSO_4 \cdot 5H_2O$
180 g anhydrous Na_2SO_4

Dilute to 1 liter with distilled H_2O (<u>Do not refrigerate</u>)

[6]Arseno-molybdate reagent: Dissolve 25 g of ammonium molybdate [$(NH_4)_6Mo_7O_{24} \cdot 4H_2O$] in 450 ml of water, add 21 ml of concentrated H_2SO_4 and mix; add 3 g of $Na_2HAsO_4 \cdot 7H_2O$ dissolved in 25 ml of water; mix; place the solution in an incubator at 37° C for 24 to 48 hours before use; store in a glass-stoppered brown bottle.

<u>sample consistently and proceed to assay duplicate samples of each in-</u>
<u>cubate filtrate</u>. (Remember to make up the volume of each sample of
filtrate to 1.0 ml, if samples less than 1.0 ml are taken, before adding
Somogyi's reagent.)

By interpolation from the glucose standard curve, determine the
glucose concentration corresponding to the absorbance value measured for
each sample of incubate filtrate (excluding those containing cycloheximide or actinomycin D), and plot, on log-log graph paper, a dosage-
response curve showing the "Reducing sugar concentration (g/ml)" (on
ordinate) versus "GA$_3$ concentration (g/ml)" (on abscissa). (<u>Remember</u>
that if the volume of incubate filtrate used in the assays was less than
1.0 ml, a correction factor must be applied to all interpolated reducing
sugar concentrations determined for the incubate filtrates, since the
sample size of each standard glucose solution was 1.0 ml.) Record in a
table the results obtained with actinomycin D and cycloheximide.

References

1. Briggs, D. E. 1963. Biochemistry of barley germination: action
    of gibberellic acid on barley endosperm. J. Inst. Brewing
    69: 13-19.

2. Chrispeels, M. J. and J. E. Varner. 1967a. Gibberellic acid-
    enhanced synthesis and release of α-amylase and ribonuclease
    by isolated barley aleurone layers. Plant Physiol. 42:
    398-406.

3. Chrispeels, M. J. and J. E. Varner. 1967b. Hormonal control of
    enzyme synthesis: on the mode of action of gibberellic
    acid and abscisin in aleurone layers of barley. Plant
    Physiol. 42: 1008-1016.

4. Cohen, D. and L. G. Paleg. 1967. Physiological effects of
    gibberellic acid. X. The release of gibberellin-like
    substances by germinating barley embryos. Plant Physiol.
    42: 1288-1296.

5. Coombe, B. G., D. Cohen, and L. G. Paleg. 1967a. Barley endo-
    sperm bioassay for gibberellins. I. Parameters of the
    response system. Plant Physiol. 42: 105-112.

6. Coombe, B. G., D. Cohen, and L. G. Paleg. 1967b. Barley endo-
    sperm bioassay for gibberellins. II. Application of the
    method. Plant Physiol. 42: 113-119.

7. Evins, W. H. and J. E. Varner. 1972. Hormonal control of
    polyribosome formation in barley aleurone layers. Plant
    Physiol. 49: 348-352.

8. Filner, P. and J. E. Varner. 1967. A test for de novo synthesis of enzymes: density labeling with H_2O^{18} of barley α-amylase induced by gibberellic acid. Proc. Nat. Acad. Sci. 58: 1520-1526.

9. Galston, A. W. and P. J. Davies. 1969. Hormonal regulation in higher plants. Science 163: 1288-1297.

10. Galston, A. W. and P. J. Davies. 1970. Control Mechanisms in Plant Development. Prentice-Hall, Englewood Cliffs, New Jersey.

11. Higgins, T. J. V., J. A. Zwar, and J. V. Jacobsen. 1976. Gibberellic acid enhances the level of translatable mRNA for α-amylase in barley aleurone layers. Nature 260: 166-169.

12. Jacobsen, J. V. and J. E. Varner. 1967. Gibberellic acid-induced synthesis of protease by isolated aleurone layers of barley. Plant Physiol. 42: 1596-1600.

13. Jarvis, B. C., B. Frankland, and J. H. Cherry. 1968. Increased nucleic acid synthesis in relation to the breaking of dormancy of hazel seed by gibberellic acid. Planta 83: 257-266.

14. Johri, M. M. and J. E. Varner. 1968. Enhancement of RNA synthesis in isolated pea nuclei by gibberellic acid. Proc. Nat. Acad. Sci. 59: 269-276.

15. Johri, M. M. and J. E. Varner. 1970. Characterization of rapidly labeled ribonucleic acid from dwarf peas. Plant Physiol. 45: 348-358.

16. Jones, K. C. 1968. Time of initiation of the barley endosperm response to gibberellin A_3, gibberellin A_{14} and kaurene. Planta 78: 366-370.

17. Jones, R. L. 1971. Gibberellic acid-enhanced release of β-1, 3-glucanase from barley aleurone cells. Plant Physiol. 47: 412-416.

18. Jones, R. L. and J. E. Varner. 1967. The bioassay of gibberellins. Planta 72: 155-167.

19. Key, J. L. 1969. Hormones and nucleic acid metabolism. Ann. Rev. Plant Physiol. 20: 449-474.

20. Lang, A. 1970. Gibberellins: structure and metabolism. Ann. Rev. Plant Physiol. 21: 537-570.

21. MacLeod, A. M. and A. S. Millar. 1962. Effect of gibberellic acid on barley endosperm. J. Inst. Brewing 68: 322-332.

22. Momotani, Y. and J. Kato. 1966. Isozymes of α-amylase induced
    by gibberellic acid in embryo-less grains of barley. Plant
    Physiol. 41: 1395-1396.

23. Moore, T. C. 1979. Biochemistry and Physiology of Plant Hormones.
    Springer-Verlag, New York.

24. Paleg, L. G. 1960. Physiological effects of gibberellic acid.
    II. On starch hydrolyzing enzymes of barley endosperm.
    Plant Physiol. 35: 902-906.

25. Paleg, L. G. 1961. Physiological effects of gibberellic acid.
    III. Observations on its mode of action on barley endosperm.
    Plant Physiol. 36: 829-837.

26. Paleg, L. G. 1965. Physiological effects of gibberellins. Ann.
    Rev. Plant Physiol. 16: 291-322.

27. Phillips, I. D. J. 1971. Introduction to the Biochemistry and
    Physiology of Plant Growth Hormones. McGraw-Hill Book
    Company, New York.

28. Poulson, R. and L. Beevers. 1970. Effects of growth regulators
    on ribonucleic acid metabolism of barley leaf segments.
    Plant Physiol. 46: 782-785.

29. Radley, M. 1967. Site of production of gibberellin-like substances
    in germinating barley embryos. Planta 75: 164-171.

30. Salisbury, F. B. and C. W. Ross. 1978. Plant Physiology. 2nd Ed.
    Wadsworth Publishing Company, Belmont, California.

31. Steward, F. C. and A. D. Krikorian. 1971. Plants, Chemicals and
    Growth. Academic Press, New York.

32. Taiz, L. and R. L. Jones. 1970. Gibberellic acid, β-1, 3-
    glucanase and the cell walls of barley aleurone layers.
    Planta 92: 73-84.

33. Varner, J. E. 1964. Gibberellic acid controlled synthesis of
    α-amylase in barley endosperm. Plant Physiol. 39: 413-415.

34. Varner, J. E. and G. Ram Chandra. 1964. Hormonal control of
    enzyme synthesis in barley endosperm. Proc. Nat. Acad. Sci.
    52: 100-106.

35. Varner, J. E., G. Ram Chandra, and M. J. Chrispeels. 1965.
    Gibberellic acid controlled synthesis of α-amylase in barley
    endosperm. J. Cellular Comp. Physiol. 66: 55-68.

36. Yomo, H. 1960. Studies on the amylase activating substance. IV.
    On the amylase activating action of gibberellin. Hakko
    Kyokashi 18: 600-602.

Special Materials and Equipment Required
Per Team of 3-8 Students

(Approximately 50) Barley (<u>Hordeum</u> <u>vulgare</u>) grains. The popular huskless
    variety Himalaya can be purchased from the Department of Agronomy,
    Washington State University, Pullman, Washington 99163. Grains two
    or three years old are preferred, since newly harvested seeds contain
    substantial levels of GA which give rise to relatively high "background"
    levels of α-amylase.
(16) Glass vials approximately 24 mm diameter and 50 mm height (e.g.
    Kimble stopper vials, 25.25 X 52 mm)
(1) Test tube rack to hold vials
(1) Mechanical shaker
(Approximately 1 liter) Sterile distilled water
(1) Sterile packet containing paper towels, forceps, 1 to several 1-ml
    pipets, razor blade or scalpel and 16 new corks (size 7) to fit
    vials
(Approximately 5 ml) Each of the following test solutions: 0, 10^{-11},
    10^{-10}, 10^{-9}, 10^{-8} and 10^{-7} g/ml GA_3; 10^{-8} g/ml GA_3 + 10 µg/ml cyclo-
    heximide; and 10^{-8} g/ml GA_3 + 100 µg/ml actinomycin D
(16) Disks Whatman No. 1 filter paper, 4.25 cm
(Approximately 75 ml) Somogyi's alkaline copper reagent, prepared as
    described in Materials and Methods
(Approximately 75 ml) Arseno-molybdate reagent, prepared as described in
    Materials and Methods
(Approximately 5 ml) Each of the following glucose solutions: 0, 10^{-7},
    10^{-6}, 10^{-5}, 3.3 x 10^{-5}, 10^{-4}, 3.3 x 10^{-4} and 10^{-3} g/ml.
(1) Pipet filler
(Approximately 16 g) Ion exchange resin, e.g. Amberlite IR120 (H^+)
(1) Colorimeter or spectrophotometer

Recommendations for Scheduling

    It is recommended that students in each laboratory section perform
this experiment in 3 to 4 teams, with each team doing the entire experiment. The GA, GA + inhibitor, and glucose solutions; Somogyi's alkaline
copper reagent; and arseno-molybdate reagent should be prepared in advance
of the periods in which they are needed. It is advisable also to soak
the sterilized barley grains in advance of the period in which the students
are to set up the experiment. For this reason, some instructors may choose
to follow the first alternative (described in a footnote) to the prescribed
procedure. It should be noted that an inhibitory effect of actinomycin D
is more readily demonstrable if the alternative procedure is followed.

<u>Duration</u>: 2 to 3 days, depending upon which alternative procedure is
followed. If the primary procedure described in the Materials and Methods
is followed, approximately 1 hour is required during the period when the
hydrated half-grains are placed in the incubation solutions, and approximately 2 hours are required during the period when the incubate filtrates
are analyzed.

EXERCISE 16

INDUCTION OF α-AMYLASE SYNTHESIS IN ALEURONE CELLS OF BARLEY GRAINS BY GIBBERELLIN

REPORT

Name _____ Section _____ Date _____

Assays of standard glucose solutions:

Concentration (g/ml)	Absorbance (O.D.) (560 nm)		
	Sample 1	Sample 2	Mean
10^{-7}	_____	_____	_____
10^{-6}	_____	_____	_____
10^{-5}	_____	_____	_____
3.3×10^{-5}	_____	_____	_____
10^{-4}	_____	_____	_____
3.3×10^{-4}	_____	_____	_____
10^{-3}	_____	_____	_____

INDUCTION OF α-AMYLASE SYNTHESIS IN ALEURONE CELLS OF BARLEY GRAINS BY GIBBERELLIN

REPORT

Name _____ Section _____ Date _____

<u>Glucose standard curve:</u>

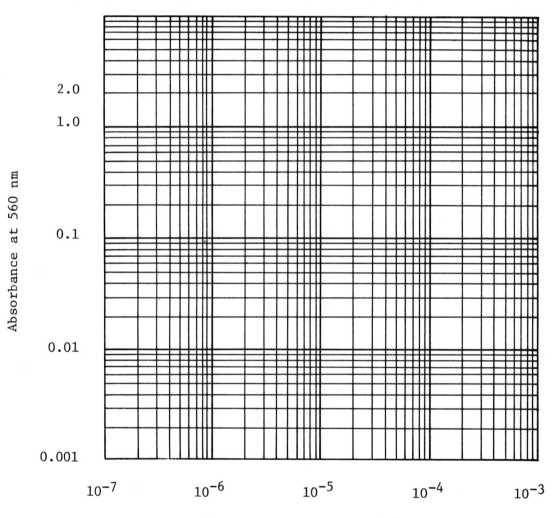

EXERCISE 16

INDUCTION OF α-AMYLASE SYNTHESIS IN ALEURONE CELLS OF BARLEY GRAINS BY GIBBERELLIN

REPORT

Name _____ Section _____ Date _____

Results of colorimetric assays:

Preliminary assay of incubate filtrate (10^{-8} g/ml GA$_3$):

Sample volume (ml)	Absorbance (O.D.) (560 nm)		
	Sample 1	Sample 2	Mean
0.10	_____	_____	_____
0.25	_____	_____	_____
0.50	_____	_____	_____
1.00	_____	_____	_____

Assays of incubate filtrates:

Incubation solution (g/ml GA$_3$)	Absorbance (O.D.) (560 nm)		
	Sample 1	Sample 2	Mean
0	_____	_____	_____
10^{-11}	_____	_____	_____
10^{-10}	_____	_____	_____
10^{-9}	_____	_____	_____
10^{-8}	_____	_____	_____
10^{-7}	_____	_____	_____
10^{-8} + CH	_____	_____	_____
10^{-8} + Act. D	_____	_____	_____

EXERCISE 16

INDUCTION OF α-AMYLASE SYNTHESIS IN ALEURONE CELLS OF BARLEY GRAINS BY GIBBERELLIN

REPORT

Name _____ Section _____ Date _____

GA₃ dosage-response curve:

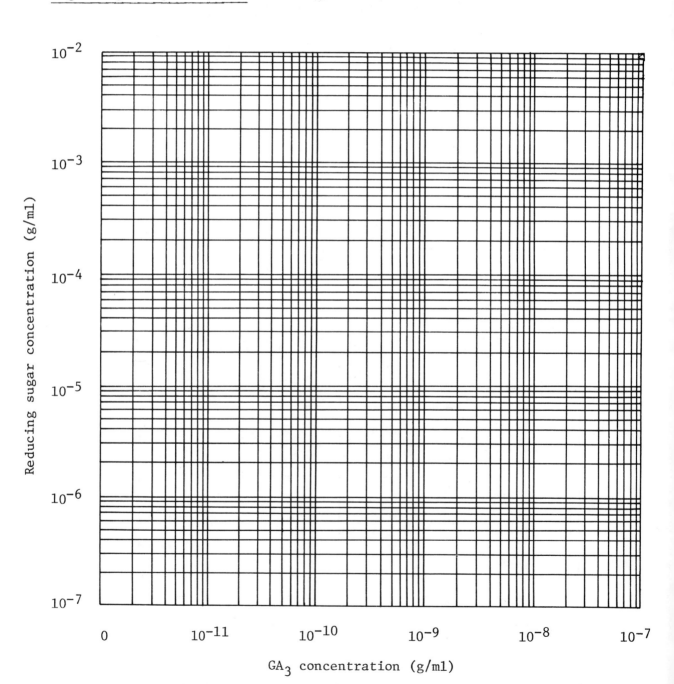

INDUCTION OF α-AMYLASE SYNTHESIS IN ALEURONE CELLS OF BARLEY GRAINS BY GIBBERELLIN

REPORT

Name _____ Section _____ Date _____

Effects of cycloheximide and actinomycin D:

Incubation solution	Reducing sugar concentration (g/ml)
GA	_____
GA + CH	_____
GA + Act. D	_____

EXERCISE 16

INDUCTION OF α-AMYLASE SYNTHESIS IN ALEURONE CELLS OF BARLEY GRAINS BY GIBBERELLIN

REPORT

Name _____ Section _____ Date _____

Questions

1. Explain why embryo-less half-grains instead of whole grains were used in this experiment.

2. Could the induction of α-amylase synthesis by GA be demonstrated with isolated aleurone layers? If so, what fundamental changes in experimental procedure would be necessary?

3. Describe, on the basis of currently available evidence, the specific sequence of events evoked by GA which culminated in release of reducing sugar from the embryo-less barley half-grains.

EXERCISE 16

INDUCTION OF α-AMYLASE SYNTHESIS IN ALEURONE CELLS OF BARLEY GRAINS BY GIBBERELLIN

REPORT

Name _____ Section _____ Date _____

4. Explain briefly the physiological importance of the GA action witnessed in this experiment in the natural germination of barley grains.

5. What <u>specific and unequivocal</u> conclusions can be made about the effects of actinomycin D and cycloheximide which were observed in this experiment?

6. Calculate the molarity of the GA_3 solution which evoked the maximum release of reducing sugar in this experiment.

Exercise 17

Effects of a Cytokinin on Bean Leaf Growth and Senescence

Senescence is one of the most poorly defined and least understood--yet very important--phenomena in biology. By a generally acceptable connotation, <u>senescence</u> refers to the collective progressive and deteriorative processes which naturally terminate in the death of an organ or organism. In this sense "senescence" is synonymous with "aging."

There are many familiar manifestations of senescence in higher plants. Whole plant senescence is obvious in annual plants in which the life cycle is terminated at the end of one growing season. In herbaceous perennials, the aerial portion of the shoot senesces and dies at the end of a growing season, but the underground system remains viable. Annual senescence and abscission of leaves is exhibited by many woody perennials. Progressive senescence and death of leaves from the basal portion acropetally on a shoot is common. Finally, the ripening and subsequent deterioration of fleshy fruits is an example of senescence.

The biochemical changes that occur in leaves as they grow old have been characterized extensively. However, the endogenous factors which regulate these changes and the reasons why the leaf eventually dies are incompletely understood. The most prominent changes that occur in the senescent leaf are declines in the proteins and nucleic acids and an irreversible yellowing due to the loss of chlorophyll. Catabolism exceeds anabolism, and there is massive export of a variety of soluble metabolites out of the senescent leaf to other parts of the plant. Photosynthesis declines, and generally respiration rate decreases also. Cytological changes include structural, as well as functional, disorganization of organelles, increased membrane permeability, destruction of cytoplasmic polyribosomes and eventual disintegration of nuclei.

Leaf senescence and abscission (in those species in which abscission occurs) actually is a conservative biological phenomenon. Lower (older) leaves on a shoot axis often become quite inefficient organs of photosynthesis because of the shading by the upper portion of the shoot. Thus death and possible elimination of the organ, preceded by export of soluble metabolites, may well be a conservative process which contributes to the overall welfare of the plant.

When a mature leaf is excised from the plant and the petiole (or basal part) submerged basally in water, senescence occurs rapidly, provided the petiole does not form adventitious roots. The protein and chlorophyll content decline to less than half the original value in a matter of just a few days. The same process can be observed with excised sections or disks of leaves floated on water. If, however, the excised leaf forms adventitious roots, its behavior is quite different. It remains green and healthy and frequently will live longer than if it had remained attached to the plant. An explanation for this behavior will be presented later in this discussion.

In 1957, A. E. Richmond and A. Lang discovered that kinetin (Fig. 17-1), one of the group of growth substances known collectively as

191

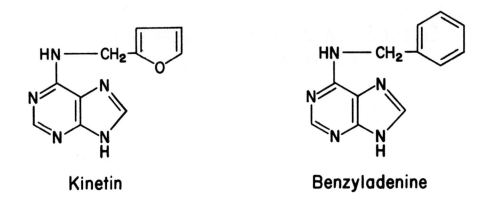

Kinetin   Benzyladenine

Fig. 17-1. Structures of two cytokinins.

cytokinins, retarded the senescence of detached leaves of cocklebur
(Xanthium). This discovery was rapidly followed by numerous reports
showing that kinetin and benzyladenine (Fig. 17-1), another synthetic
cytokinin, retard senescence of whole excised leaves and excised sections
of leaves of numerous species. If the leaf sections are quite young when
excised, cytokinin stimulates growth appreciably.

   This area of research was advanced dramatically in 1959 and the
years following by K. Mothes and associates in Germany. They found by
spraying solutions of kinetin directly onto leaves that the effect of
the applied cytokinin was quite localized. Only those areas to which
the chemical was applied remained green. Furthermore, the treated areas
of yellowing leaves actually became greener. In an effort to explain
these intriguing observations, Mothes and associates applied radioactive
amino acids and non-radioactive kinetin to leaves. They found that
radioactive amino acids migrated to and accumulated in the kinetin-treated
parts of the leaves. Thus it came to be understood that cytokinin causes
mobilization of metabolites from an untreated to a treated portion of a
leaf, and that the chemical somehow acts to sustain nucleic acid and
protein synthesis.

   A. C. Leopold and M. Kawase (1964) advanced this research a step
further using shoot cuttings of bean (Phaseolus vulgaris) seedlings.
They observed another interesting effect of benzyladenine in addition to
stimulation of growth of intact whole young leaves and retardation of
senescence of localized treated parts of mature leaves. Treatment of
one leaf with cytokinin actually caused inhibition of growth and hastened
senescence of other, untreated leaves on the cuttings.

   Previously it was stated that if adventitious roots form on the
petiole of an excised leaf rapid senescence, typical of the excised mature
leaf, does not occur. An explanation for this behavior began to emerge
in 1964 and 1965, with reports by H. Kende and by C. Itai and Y. Vaadia
that roots apparently export endogenous cytokinin in the xylem sap.

The purpose of this experiment is to investigate the effects of benzyladenine on leaf growth and senescence on shoot cuttings of bean seedlings.

Materials and Methods[1,2]

Approximately 100 to 150 seeds of bush bean or Red Kidney bean (Phaseolus vulgaris) were planted previously in each of two flats filled with vermiculite for each laboratory section. The plants were grown in a greenhouse or growth chamber under relatively long photoperiods (approximately 16 hours), moderately high light intensity (1500 to 2000 ft-c), and a day-night temperature cycle of approximately 24° to 20° C. The plants were irrigated as needed with tap water (not mineral nutrient solution).

When the plants are approximately 14 days old; that is, when they have attained a shoot length of approximately 25 cm and have a pair of mature primary leaves and an immature but macroscopic first trifoliate leaf, harvest the cuttings by excising the shoots approximately 10 cm below the paired primary leaves. Quickly remove the cotyledons from the cuttings, and put 5 to 8 cuttings in each of 5 large beakers (400-ml to 1-liter size) approximately half-filled with distilled water. Be sure the base of each cutting is submerged!

Number the beakers 1 through 5, and treat the cuttings as follows:

Group No. 1 – Cut off the existing trifoliate leaf and other trifoliate leaves as they form, leaving the shoot tip

---

[1]Adopted from: Leopold, A. C. and M. Kawase. 1964. Benzyladenine effects on bean leaf growth and senescence. Am. J. Botany 51: 294-298.

[2]A satisfactory and somewhat simplified alternative to this procedure can be adapted from: Osborne, D. J. 1962. Effect of kinetin on protein and nucleic acid metabolism in Xanthium leaves during senescence. Plant Physiol. 37: 595-602. Leaf discs approximately 1.5 cm in diameter are cut with a cork borer from interveinal areas of the blades of mature primary leaves of bean plants (or other species). Groups of discs are floated on distilled water or benzyladenine solution (approximately 40 mg/liter) in petri dishes in darkness at about 24° C. The senescence-delaying effect of cytokinin is readily apparent after 3 to 4 days incubation. Results are quantified by a crude chlorophyll extraction procedure (described above), and either time of incubation (days) or cytokinin concentration is used as the variable.

intact. Apply benzyladenine (BA) solution[3] to one primary leaf and leave the other primary leaf untreated.

Group No. 2 – Same as for Group No. 1, except treat half (on one side of midrib) of each primary leaf; leave opposite half of each leaf untreated.

Group No. 3 – Apply BA to both primary leaves and leave the intact trifoliate leaves untreated.

Group No. 4 – Apply BA to the leaflets of the first trifoliate leaf and leave the primary leaves untreated.

Group No. 5 – Remove no leaves; apply no BA.

Set the cuttings in a growth chamber or elsewhere under constant white light (preferably mixed fluorescent and incandescent light) and a constant temperature of approximately 22 to 24° C.

Twice a week make a fresh cut at the base of each cutting, change the distilled water, remove any adventitious roots that are present, and repeat the treatment with BA.

Approximately 14 days after the first treatment, terminate the experiment. Sketch one plant from each group, using green and yellow pencils to illustrate the coloration of the leaves. Measure the lengths of the lamina of individual leaflets of the first trifoliate leaf of all cuttings in Groups 3, 4 and 5; record the mean values in a table. Finally, harvest samples of 5 whole leaf blades from each of the pairs of primary leaves on the cuttings in Groups 1, 3 and 5. Extract the chloroplast pigments from each of the 6 samples separately by grinding with a few ml of acetone in a mortar and pestle. Centrifuge or filter the extracts to remove particulate matter, dilute the supernatants with acetone to a uniform final volume,[4] and measure the % absorption of light at 665 nm with a colorimeter or spectrophotometer. Present the absorption data in a table.

---

[3]Prepare 500 ml of a solution of benzyladenine (BA) at a concentration of 30 mg/l by dissolving 15 mg of BA in 1 to 2 ml of 0.3 N KOH and diluting to 500 ml with distilled water containing 0.05% Tween 20.
  Mark the leaves or portions of leaves to be treated with a "T" written gently with India ink or waterproof felt pen.
  Apply the BA solution with a soft brush or cotton swab (gently) on the upper (adaxial) side of the leaf.

[4]Adjust the volume of the most optically dense (greenest) extract so that it absorbs less than 100% at 665 nm; dilute all other extracts to the same final volume.

References

1. Beevers, L. 1966. Effect of gibberellic acid on the senescence of leaf discs of Nasturtium (Tropaeolum majus). Plant Physiol. 41: 1074-1076.

2. Ecklund, P. R. and T. C. Moore. 1968. Quantitative changes in gibberellin and RNA correlated with senescence of the shoot apex in the 'Alaska' pea. Am. J. Botany 55: 494-503.

3. Fletcher, R. A. and D. J. Osborne. 1966. Gibberellin, as a regulator of protein and ribonucleic acid synthesis during senescence in leaf cells of Taraxacum officinale. Can. J. Botany 44: 739-745.

4. Fox, J. E. 1969. The cytokinins. Pp. 85-123 in: M. B. Wilkins, Ed. The Physiology of Plant Growth and Development. McGraw-Hill Publishing Company, New York.

5. Galston, A. W. and P. J. Davies. 1970. Control Mechanisms in Plant Development. Prentice-Hall, Englewood Cliffs, New Jersey.

6. Itai, C. and Y. Vaadia. 1965. Kinetin-like activity in root exudates of water-stressed sunflower plants. Physiol. Plantarum 18: 941-944.

7. Kende, H. 1964. Preservation of chlorophyll in leaf sections by substances obtained from root exudate. Science 145: 1066-1067.

8. Kende, H. 1965. Kinetin-like factors in the root exudate of sunflowers. Proc. Nat. Acad. Sci. 53: 1302-1307.

9. Krizek, D. T., W. J. McIlrath, and B. S. Vergara. 1966. Photoperiodic induction of senescence in Xanthium plants. Science 151: 95-96.

10. Leopold, A. C. and M. Kawase. 1964. Benzyladenine effects on bean leaf growth and senescence. Am. J. Botany 51: 294-298.

11. Leopold, A. C. and P. E. Kriedemann. 1975. Plant Growth and Development. 2nd Ed. McGraw-Hill Book Company, New York.

12. Leopold, A. C., E. Niedergang-Kamien, and J. Janick. 1959. Experimental modification of plant senescence. Plant Physiol. 34: 570-573.

13. Letham, D. S. 1967a. Chemistry and physiology of kinetin-like compounds. Ann. Rev. Plant Physiol. 18: 349-364.

14. Letham, D. S. 1967b. Regulators of cell division in plant tissues. V. A comparison of the activities of zeatin and other cytokinins in five bioassays. Planta 74: 228-242.

15. Lockhart, J. A. and V. Gottschall. 1961. Fruit-induced and apical senescence in Pisum sativum L. Plant Physiol. 36: 389-398.

16. Moore, T. C. 1979. Biochemistry and Physiology of Plant Hormones. Springer-Verlag, New York.

17. Mothes, K. and L. Engelbrecht. 1961. Kinetin-induced directed transport of substances in excised leaves in the dark. Phytochemistry 1: 58-62.

18. Osborne, D. J. 1959. Control of leaf senescence by auxins. Nature 183: 1459-1460.

19. Osborne, D. J. 1962. Effect of kinetin on protein and nucleic acid metabolism in Xanthium leaves during senescence. Plant Physiol. 37: 595-602.

20. Person, C., D. J. Samborski, and F. R. Forsyth. 1957. Effect of benzimidazole on detached wheat leaves. Nature 180: 1294-1295.

21. Phillips, I. D. J. 1971. Introduction to the Biochemistry and Physiology of Plant Growth Hormones. McGraw-Hill Book Company, New York.

22. Richmond, A. E. and A. Lang. 1957. Effect of kinetin on protein content and survival of detached Xanthium leaves. Science 125: 650-651.

23. Sacher, J. A. 1965. Senescence: hormonal control of RNA and protein synthesis in excised bean pod tissue. Am. J. Botany 52: 841-848.

24. Sax, K. 1962. Aspects of aging in plants. Ann. Rev. Plant Physiol. 13: 489-506.

25. Seth, A. K. and P. F. Wareing. 1967. Hormone-directed transport of metabolites and its possible role in plant senescence. J. Exp. Bot. 18: 65-77.

26. Shibaoka, H. and K. V. Thimann. 1970. Antagonisms between kinetin and amino acids. Experiments on the mode of action of cytokinins. Plant Physiol. 46: 212-220.

27. Shaw, M., P. K. Bhattacharya, and W. A. Quick. 1965. Chlorophyll, protein, and nucleic acid levels in detached, senescing wheat leaves. Can. J. Botany 43: 739-746.

28. Skoog, F. and D. J. Armstrong. 1970. Cytokinins. Ann. Rev. Plant Physiol. 21: 359-384.

29. Sorokin, C. 1964. Aging at the cellular level. Experientia
    20: 1-10.

30. Steward, F. C. and A. D. Krikorian. 1971. Plants, Chemicals
    and Growth. Academic Press, New York.

31. Tavares, J. and H. Kende. 1970. The effect of 6-benzylamino-
    purine on protein metabolism in senescing corn leaves.
    Phytochemistry 9: 1763-1770.

32. Varner, J. E. 1961. Biochemistry of senescence. Ann. Rev.
    Plant Physiol. 12: 245-264.

33. Wareing, P. F. and I. D. J. Phillips. 1970. The Control of
    Growth and Differentiation in Plants. Pergamon Press,
    New York.

34. Whyte, P. and L. C. Luckwill. 1966. A sensitive bioassay for
    gibberellins based on retardation of leaf senescence in
    Rumex obtusifolius (L.). Nature 210: 1360.

35. Woolhouse, H. W., Ed. 1967. Aspects of the biology of aging.
    Symposia of the Society for Experimental Biology, No. 21.
    Academic Press, New York.

Special Materials and Equipment Required
Per Team of 3-8 Students

(25 to 50) Bush bean or Red Kidney bean (Phaseolus vulgaris) seedlings
    approximately 14 days old; the plants should be about 25 cm tall
    and have mature primary leaves, and the leaflets of the first tri-
    foliate leaf should be about 1 cm long
(Approximately 100 ml) BA solution, prepared as described in Materials
    and Methods
(1) India ink or waterproof felt marking pen
(1 to several) Cotton swabs or small, soft brushes
(Approximately 100 ml) Acetone
(1) Colorimeter or spectrophotometer

Recommendations for Scheduling

This experiment can be done quite satisfactorily by teams of 3 to 8
students, each team doing the entire experiment. The bean plants should
be grown to appropriate size in advance of the laboratory period when the
experiment is to be started. The BA solution may also be prepared in
advance.

Duration: Approximately 14 days; 30 to 45 minutes required during
first laboratory period, 15 minutes required at least twice a week thereafter until the final period, and approximately 1 to 1.5 hours required
during the period when the experiment is terminated.

EXERCISE 17

EFFECTS OF A CYTOKININ ON BEAN LEAF
GROWTH AND SENESCENCE

REPORT

Name _____ Section _____ Date _____

Sketches of cuttings:

EFFECTS OF A CYTOKININ ON BEAN LEAF
GROWTH AND SENESCENCE

REPORT

Name _____ Section _____ Date _____

Measurements of lengths of individual leaflet lamina of the first
trifoliate leaf in Groups 3, 4 and 5:

Group no.	Treatment	Mean leaf length (cm)
3	BA on both primary leaves; trifoliate leaf untreated	_____
4	BA on trifoliate leaf; primary leaves untreated	_____
5	No BA treatment	_____

Measurements of % absorption of light at 665 nm in acetone extracts of
primary leaves in Groups 1, 3 and 5:

Group no.	Treatment	Leaf Sample	Treated with BA	% Absorption (665 nm)
1	BA on one primary leaf; trifoliate leaf removed	1	−	_____
		2	+	_____
3	BA on both primary leaves; trifoliate leaf intact	1	+	_____
		2	+	_____
5	No BA treatment; trifoliate leaf intact	1	−	_____
		2	−	_____

EXERCISE 17

EFFECTS OF A CYTOKININ ON BEAN LEAF
GROWTH AND SENESCENCE

REPORT

Name _____ Section _____ Date _____

Questions

1. Summarize briefly the results of treatment with benzyladenine of each
    of the Groups 1 through 4, as compared to the cuttings in Group 5.

2. What specific evidence, if any, is there from this experiment that
    treatment of one leaf actually hastened senescence of another, un-
    treated leaf (or leaves) on the same plant?

3. Did benzyladenine promote the growth of any of the treated leaves or
    leaflets? Discuss.

EFFECTS OF A CYTOKININ ON BEAN LEAF
GROWTH AND SENESCENCE

REPORT

Name _____ Section _____ Date _____

4. Why were adventitious roots removed from the cuttings?

5. Describe some practical applications, based on the senescence-retarding action of cytokinins, which have been developed in floriculture and horticulture.

6. What is benzimidazole? Is it a cytokinin?

EFFECTS OF ETHYLENE ON THE DEVELOPMENT
OF LEGUME SEEDLINGS

Introduction

Ethylene ($H_2C=CH_2$) is now almost universally regarded as a plant hormone. Recognition of this relatively simple hydrocarbon as a hormone came about reluctantly on the part of many physiologists, however, even after the compound was recognized as a natural plant product which influences growth and development in numerous ways. The reason for the hesitation was that ethylene is unique among endogenous growth regulators in being a vapor at physiological temperatures. Thus the metabolism, transport and action of ethylene possess features and involve problems not commonly encountered in hormone biochemistry and physiology.

The history of ethylene physiology dates back to early investigations of ripening and postharvest physiology of fleshy fruits. In the two decades from 1917 to 1937, it was discovered that ethylene stimulates fruits to ripen, and that indeed ripening fruits, as well as some other plant materials, produce the gas. These observations led W. Crocker, A. E. Hitchcock and P. W. Zimmerman of the Boyce Thompson Institute for Plant Research at Yonkers, New York to suggest in 1935 that ethylene be regarded as a ripening hormone.

For many years ethylene physiology remained primarily the province of those interested in fruit physiology. Several practical applications developed from this research, among them being the process of "gas storage" for the preservation of fruits, which was developed by Kidd and West in 1934. It had been determined that oxygen is necessary for ethylene action and that carbon dioxide inhibits action of the gas. Basically the process of "gas storage" consists of storing fruits in an atmosphere rich in carbon dioxide (5 to 10%), low in oxygen (1 to 3%) and with as little ethylene as possible. Gas storage is achieved commercially by storing fruits in an airtight room at low temperature. The oxygen is naturally depleted and the carbon dioxide increased by respiration. Ethylene produced by the fruits is absorbed on brominated charcoal filters. In the past few years the carbon dioxide effect has been explained by the discovery that carbon dioxide is a competitive inhibitor of ethylene action.

Resurgence and great expansion of interest in ethylene physiology occurred in the late 1950's, and this has been a very active and productive field of research since that time. Much of the expansion of research in this field can be directly attributed to the advent of gas chromatography. Until this analytical instrumentation became available, it simply was not possible to measure the very low concentrations of ethylene present in many plant tissues. With the best flame-ionization gas chromatographs it is now possible to detect and measure less than one part ethylene per thousand million in a small gas sample. Particularly prominent in ethylene physiology in recent years have been Stanley P. and Ellen A. Burg at the University of Miami School of Medicine, and Harlan K. Pratt and associates at the University of California at Davis.

Ethylene now is understood to be a natural product of plant metabolism, which is produced by healthy (as well as senescent and diseased) tissues and which exerts regulatory control or influence over plant growth and development. Among the many diverse physiological effects of ethylene which have been reported are the following:

1. Stimulation of ripening of fleshy fruits.
2. Stimulation of leaf abscission.
3. "Triple response" of etiolated legume seedlings--reduced stem elongation, increased radial swelling of stem, and transverse geotropism (horizontal orientation).
4. Inhibition of leaf and terminal bud expansion in etiolated seedlings.
5. Tightening of the epicotyl or hypocotyl hook of etiolated dicot seedlings.
6. Inhibition of root growth.
7. Increase in membrane permeability.
8. Stimulation of adventitious root formation.
9. Stimulation of flowering in pineapple.
10. Inhibition of lateral bud development.
11. Causes various types of flowers to fade.
12. Interference with polar auxin transport.
13. Causes epinasty of leaves.
14. Participates in normal root geotropism.

The purpose of this experiment is to investigate the effects of exogenous ethylene on the growth and development of etiolated pea (_Pisum sativum_) seedlings.

Materials and Methods[1]

The procedure consists of growing Alaska pea seedlings in gas-tight chambers which can be injected with ethylene, under carefully controlled conditions.

Prepare 6 battery jars of approximately 10- to 20-liter capacity[2] as culture chambers by first fitting each jar with a plastic or plexiglass cover which is tapped and fitted with a rubber serum cap (approximately 1.5 cm diameter). Determine the volume of one of the uniform jars by filling with water and decanting into a 1-liter graduated cylinder.

The seedlings are to be planted in plastic pots (approximately 6 inches

---

[1]Adapted from: Goeschl, J. D. and H. K. Pratt. 1968. Regulatory roles of ethylene in the etiolated growth habit of _Pisum sativum_. Pp. 1229-1242 in: F. Wightman and G. Setterfield, Eds. Biochemistry and Physiology of Plant Growth Substances. Runge Press, Ottawa.

[2]Bell jars or other containers can readily be adapted for use if battery jars are unavailable.

in diameter at the top and 4 3/4 inches in height). Plug the holes in one of the pots and measure the volume required to fill the pot to within 1 inch of the top.

Thoroughly disinfect all 6 glass jars, lids and plastic pots by washing with hot water and detergent, rinsing several times with sterile distilled water, and finally rinsing with 70 or 95% ethanol. Fill each pot to within 1 inch of the top with sterile vermiculite, and saturate the vermiculite with sterile distilled water.

Pour approximately 120 Alaska pea seeds into a clean 500-ml Erlenmeyer flask, add 150 to 200 ml of 0.5% sodium hypochlorite ($\sim$ 10% Clorox or Purex household bleach prepared with sterile distilled water), stopper the flask, and wash the seeds vigorously for 5 to 10 minutes. Then rinse the seeds 5 times with sterile distilled water. Using a sterile forceps, dipped in ethanol, plant about 15 to 20 seeds in each pot. Set one pot in each battery jar, and pour 500 ml of sterile distilled water into the bottom of each jar around the base of the plastic pot. Set the lid loosely on each container, but do not seal airtight at this time. Set the assembled culture chambers in a light-tight enclosure or darkroom at about 25° C.

On about the third to fifth day after planting, prepare to inject ethylene into 4 of the 6 containers. First, calculate the approximate gas volume in each chamber:

$$\text{Net gas volume} = \text{Volume of battery jar} - (\text{volume of plastic pot} + 500 \text{ ml})$$

Then calculate the volume (in microliters) of ethylene required to give concentrations of 0.2 and 0.8 ppm of ethylene in the gas phase of the culture chambers.

All handling of the culture chambers at this stage must be done under a safelight which will not cause de-etiolation.[3] (Apply a generous layer of anhydrous lanolin around the rim of each jar, and fasten the lid tightly with strips of tape. The lid must form a gas-tight seal.) Attach a piece of rubber tubing to the outlet on the regulator of a cylinder of compressed ethylene. Close off the open end of the tubing with a Hoffman or pinch clamp, and fill the tubing with ethylene by gradually opening the valve. Insert the needle of a 50 μl Hamilton syringe into the tubing and fill it with the necessary volume of ethylene. Inject two chambers each with sufficient ethylene to yield concentrations of 0.2 and 0.8 ppm. Leave the other two chambers as controls (no ethylene).

Leave 3 of the chambers (one of each pair) in darkness for an additional 48 hours or until the next laboratory period. Transfer the other chamber of each pair to a growth chamber or elsewhere under a constant light intensity of 200 to 500 ft-c for the same period of time. Then open

---

[3]The author uses a desk lamp fitted with a General Electric F15T8·G·6 15-watt Green-Photo fluorescent tube and covered with 8 layers of amber and 3 layers of green DuPont cellophane.

all the containers and analyze the results. Sketch to scale a representative seedling in each group; measure the shoot lengths (from cotyledonary node to terminal bud) of the plants in each group; and carefully describe the general morphology of the seedlings in each treated group, as compared to the controls.[4]

References

1. Abeles, F. B. 1967. Mechanism of action of abscission accelerators. Physiol. Plantarum 20: 442-454.

2. Abeles, F. B. 1972. Biosynthesis and mechanism of action of ethylene. Ann. Rev. Plant Physiol. 23: 259-292.

3. Andreae, W. A., M. A. Venis, F. Jursic, and T. Dumas. 1968. Does ethylene mediate root growth inhibition by indole-3-acetic acid? Plant Physiol. 43: 1375-1379.

4. Burg, S. P. 1962. The physiology of ethylene formation. Ann. Rev. Plant Physiol. 13: 265-302.

5. Burg, S. P. 1968. Ethylene, plant senescence and abscission. Plant Physiol. 43: 1503-1511.

6. Burg, S. P. and E. A. Burg. 1965. Ethylene action and the ripening of fruits. Science 148: 1190-1196.

7. Burg, S. P. and E. A. Burg. 1966a. The interaction between auxin and ethylene and its role in plant growth. Proc. Nat. Acad. Sci. 55: 262-269.

8. Burg, S. P. and E. A. Burg. 1966b. Auxin-induced ethylene formation: its relation to flowering in the pineapple. Science 152: 1269.

9. Burg, S. P. and E. A. Burg. 1967. Inhibition of polar auxin transport by ethylene. Plant Physiol. 42: 1224-1228.

10. Burg, S. P. and E. A. Burg. 1968. Ethylene formation in pea seedlings: its relation to the inhibition of bud growth caused by indole-3-acetic acid. Plant Physiol. 43: 1069-1074.

11. Chadwick, A. V. and S. P. Burg. 1967. An explanation of the inhibition of root growth caused by indole-3-acetic acid. Plant Physiol. 42: 415-420.

---

[4]An interesting optional addition to this experiment is to set up two culture chambers identical to the two control chambers and to suspend a surface-disinfected apple in each.

12. Chadwick, A. V. and S. P. Burg. 1970. Regulation of root growth by auxin-ethylene interaction. Plant Physiol. 45: 192-200.

13. Galston, A. W. and P. J. Davies. 1969. Hormonal regulation in higher plants. Science 163: 1288-1297.

14. Galston, A. W. and P. J. Davies. 1970. Control Mechanisms in Plant Development. Prentice-Hall, Englewood Cliffs, New Jersey.

15. Goeschl, J. D. and H. K. Pratt. 1968. Regulatory roles of ethylene in the etiolated growth habit of _Pisum sativum_. Pp. 1229-1242 in: F. Wightman and G. Setterfield, Eds. Biochemistry and Physiology of Plant Growth Substances. Runge Press, Ottawa.

16. Goeschl, J. D., H. K. Pratt, and B. A. Bonner. 1967. An effect of light on the production of ethylene and the growth of the plumular portion of etiolated pea seedlings. Plant Physiol. 42: 1077-1080.

17. Kang, B. G., C. S. Yokum, S. P. Burg, and P. M. Ray. 1967. Ethylene and carbon dioxide: mediation of hypocotyl hook-opening response. Science 156: 958-959.

18. Leopold, A. C. and P. E. Kriedemann. 1975. Plant Growth and Development. 2nd Ed. McGraw-Hill Book Company, New York.

19. Moore, T. C. 1979. Biochemistry and Physiology of Plant Hormones. Springer-Verlag, New York.

20. Morgan, P. W. 1969. Stimulation of ethylene evolution and abscission in cotton by 2-chloroethanephosphonic acid. Plant Physiol. 44: 337-341.

21. Phillips, I. D. J. 1971. Introduction to the Biochemistry and Physiology of Plant Growth Hormones. McGraw-Hill Book Company, New York.

22. Pratt, H. K. and J. D. Goeschl. 1969. Physiological roles of ethylene in plants. Ann. Rev. Plant Physiol. 20: 541-584.

23. Rubinstein, B. and F. B. Abeles. 1965. Relationship between ethylene evolution and abscission. Botan. Gaz. 126: 255-259.

24. Salisbury, F. B. and C. W. Ross. 1978. Plant Physiology. 2nd Ed. Wadsworth Publishing Company, Belmont, California.

25. Steward, F. C. and A. D. Krikorian. 1971. Plants, Chemicals and Growth. Academic Press, New York.

26. Zimmerman, P. W. and F. Wilcoxon. 1935. Several chemical growth substances which cause initiation of roots and other responses in plants. Contrib. Boyce Thompson Inst. 7: 209-229.

Special Materials and Equipment Required
Per Laboratory Section

(Approximately 120) Seeds of Alaska pea (Pisum sativum)
(6) 6-inch plastic pots
(6 liters) Vermiculite (autoclaved)
(6) Battery jars or other suitable glass containers of about 10- to
    20-liter volume
(6) Plastic or plexiglass lids to fit glass containers, each perforated
    and fitted with a rubber serum cap
(Approximately 1 liter) 70 or 95% ethanol
(200 ml) 0.5% sodium hypochlorite (approximately 10% commercial
    household bleach)
(Approximately 6 liters) Sterile distilled water
(1) Cylinder of compressed ethylene
(1) Hamilton syringe, e.g. 50-μl capacity
(1) Darkroom or light-tight cabinet
(1) Facility for growing plants in constant light at an intensity of
    200 to 500 ft-c
(1) Safelight, e.g. as described in Materials and Methods
(Several grams) Anhydrous lanolin

Recommendations for Scheduling

It is recommended that this experiment be conducted as a class
project in each laboratory section. Some instructors may choose to have
each laboratory section do only half the experiment, involving either
etiolated or green seedlings, and make the collective results available
to all students in the course.

Duration: 6 to 12 days, approximately 1.5 hours required during the
period when the experiment is started, 30 minutes required on the day the
ethylene is injected, and about 1.5 hours required during the period when
the experiment is terminated.

EXERCISE 18

EFFECTS OF ETHYLENE ON THE DEVELOPMENT
OF LEGUME SEEDLINGS

REPORT

Name _____ Section _____ Date _____

Sketches of seedlings at time of harvest:

EXERCISE 18

EFFECTS OF ETHYLENE ON THE DEVELOPMENT
OF LEGUME SEEDLINGS

REPORT

Name _____ Section _____ Date _____

Shoot lengths of seedlings at time of harvest:

Treatment	No. of plants	Average shoot length (cm)
No C_2H_4 – darkness ____ hr.	_____	_____
0.2 ppm C_2H_4 – darkness ____ hr.	_____	_____
0.8 ppm C_2H_4 – darkness ____ hr.	_____	_____
No C_2H_4 – light ____ hr.	_____	_____
0.2 ppm C_2H_4 – light ____ hr.	_____	_____
0.8 ppm C_2H_4 – light ____ hr.	_____	_____
Apple – darkness ____ hr.	_____	_____
Apple – light ____ hr.	_____	_____

Descriptions of morphological effects:

Ethylene treatment in darkness:

EXERCISE 18

EFFECTS OF ETHYLENE ON THE DEVELOPMENT
OF LEGUME SEEDLINGS

REPORT

Name _____ Section _____ Date _____

Ethylene treatment in light:

Apple substituted for C_2H_4:

EFFECTS OF ETHYLENE ON THE DEVELOPMENT
OF LEGUME SEEDLINGS

REPORT

Name _____ Section _____ Date _____

Questions

1. What is the "triple response" of etiolated pea seedlings to ethylene? Was the "triple response" observed in this experiment? Discuss.

2. Describe any other effects of ethylene, besides the "triple response," which were observed.

3. Describe how light affected the response of the seedlings to ethylene.

EXERCISE 18

EFFECTS OF ETHYLENE ON THE DEVELOPMENT
OF LEGUME SEEDLINGS

REPORT

Name _____ Section _____ Date _____

4. Discuss the significance of the epicotyl hook in the normal germination of pea seedlings, and the changes involving ethylene which occur
   as a pea seedling normally emerges from the soil and which account for
   straightening of the epicotyl hook.

5. If the option of including an apple, instead of ethylene, in one or
   more chambers was used, describe the similarities in appearance between
   those seedlings and those treated with exogenous ethylene. How would
   you determine whether the apples produced ethylene?

EXERCISE 19

EFFECTS OF CERTAIN SYNTHETIC PLANT GROWTH REGULATORS ON THE DEVELOPMENT OF SELECTED SPECIES

Introduction

Modification of growth and development of esthetically and economically important plants and selective elimination of undesirable species by the application of chemicals already is a demonstrated successful practical application of plant physiology. The potential for future developments in this field is nothing short of tremendous. Naturally it is of utmost importance to give close attention to prevention of pollution of the environment with toxic compounds. But even with this obviously necessary and desirable restriction, rapid progress can be expected in the use of chemicals to control plants.

Broadly speaking, a plant growth regulator may be defined as any organic chemical which is active at low concentrations (less than approximately 1 to 10 mM) in promoting, inhibiting or otherwise modifying growth and development. Defined in this way, the number of compounds classifiable as growth regulators is in the hundreds and the chemical diversity among them very extensive. Diverse as they are, growth regulators may be classified arbitrarily in four major groups: naturally-occurring growth substances or hormones; synthetic hormone-like substances; synthetic growth retardants; and synthetic herbicides.

The first group is composed of the natural auxins, gibberellins, cytokinins, a variety of growth inhibitors and ethylene. Thus all hormones are (naturally-occurring) growth regulators, although, of course, not all growth regulators are hormones. There are both naturally-occurring and purely synthetic auxins and cytokinins, for example. All are growth regulators but only the naturally-occurring compounds are hormones. 2,4-Dichlorophenoxyacetic acid (2,4-D) (Fig. 19-1) is a synthetic auxin; benzyladenine is an example of the many known synthetic cytokinins.

Besides the hormones or hormone-like growth substances, there is a group of purely synthetic growth regulators which collectively are termed growth retardants. These compounds, although chemically very diverse, possess the common property of "retarding" the growth and development of numerous species of seed plants without causing severe morphological abnormalities. Examples of this group are Alar, CCC, Amo-1618 (Fig. 19-1). Some of the growth retardants are effective on a very broad spectrum of species of seed plants, while others are effective on only a restricted group of species.

Maleic hydrazide (MH) is a synthetic growth inhibitor (Fig. 19-1). These are substances which inhibit growth in both shoot and root cells and have no stimulatory range of concentrations. In contrast to the growth retardants, MH, at sufficiently high concentrations, suppresses apical dominance by inhibiting cell division in the shoot apical meristem. Like 2,4-D, MH has been investigated for many years and has been utilized extensively on agronomic and other plants.

215

A fourth group of growth regulators is the <u>herbicides</u>. Of particular interest and utility are those herbicides which exhibit selective toxicity toward only certain species. The oldest and best known of these is 2,4-D, which is an <u>auxin-type herbicide</u>. Whereas 2,4-D is perhaps best known for its capacity to kill plants, at sufficiently low concentrations it promotes growth in a manner apparently identical to indoleacetic acid, the major natural auxin. Other auxin-type herbicides are 2,4,5-T (2,4,5-trichlorophenoxyacetic acid), dicamba (2-methyl-3,6-dichlorobenzoic acid), and picloram (4-amino-3,5-trichloropicolinic acid).

In addition to the auxin-type compounds, there are <u>other herbicides</u> (Fig. 19-2), which may be classified as: <u>chlorinated aliphatic acids</u> [e.g. dalapon (2,2-dichloropropionic acid)]; <u>substituted dinitrophenols</u> [e.g. DNBP (4,6-dinitro-o-<u>sec</u>-butylphenol)]; <u>triazines</u> [e.g. atrazine (2-chloro-4-ethylamino-6-isopropylamino-1,3,5-triazine)]; <u>substituted ureas</u> [e.g. monuron 3(4-chlorophenyl)1,1-dimethyl urea]; <u>carbamates</u> [e.g. IPC (isopropyl N-phenylcarbamate)]; <u>thiocarbamates</u> [e.g. EPTC (ethyl N,N-dipropylthiocarbamate)]; <u>substituted uracils</u> [e.g. Bromacil (5-bromo-3-<u>sec</u>-butyl-6-methyluracil)]; and several other types.

The judicious and effective utilization of available growth regulators necessitates understanding the basis for the observed physiological effects and the selective action, if any, of the compounds. Thus, prominent among the features of each compound which are commonly investigated are its absorption, translocation, metabolism, and mechanism or mode of action, as well as its toxicity to animals and persistence and movement in the environment.

The purpose of this experiment (or series of related experiments) is to investigate the effects of certain synthetic growth-regulating chemicals (Fig. 19-1) on the development of selected monocot and dicot species.

Materials and Methods

Seedlings of several species, approximately 9 to 14 days of age and potted in either soil or vermiculite, will be available for treatment. The experiment will be conducted as a class project in the greenhouse. Each student will prepare an independent written report on the entire experiment. The specific growth regulators to be used, the kinds of plants to be treated with each regulator, and the procedures to be followed are described below.

A. <u>Maleic hydrazide (MH) (Mol. Wt. 112)</u>. Using an atomizer sprayer, treat one group each of bush bean and corn seedlings with an aqueous 4 X 10^{-2} M solution (=4,480 mg/1) of MH[1] containing 0.05% Tween 20. Spray one group of each kind of plant also with 0.05% Tween 20 only.

---

[1]Dissolve 1.12 g of MH in 250 ml of distilled water (heat to approximately 50° C and stir continuously for 1 to 2 hours); then add approximately 0.1 to 0.2 ml of Tween 20 and mix thoroughly.

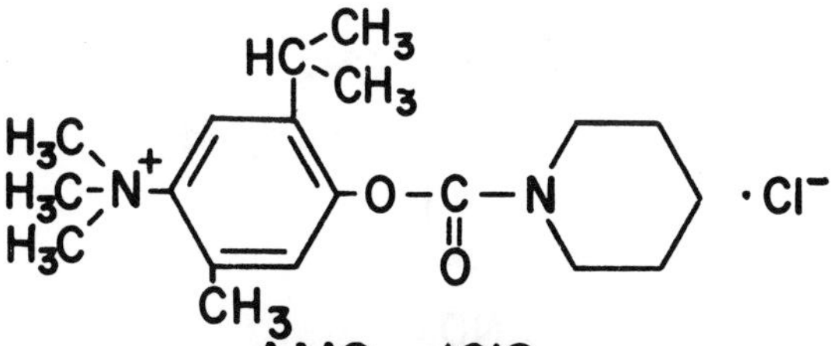

HC$\cdot$CH$_3$ / CH$_3$

H$_3$C, H$_3$C—N$^+$, H$_3$C — (phenyl ring) — O—C(=O)—N (piperidine) · Cl$^-$

CH$_3$

AMO – 1618
(4-Hydroxy-5-isopropyl-2-methyl
phenyl)-trimethylammonium chloride
1-piperidine carboxylate

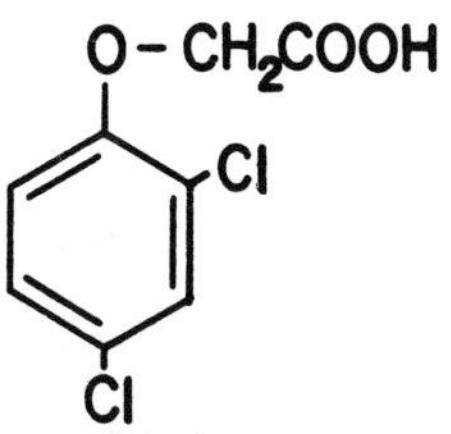

O—CH$_2$COOH
Cl
Cl

2,4-D
2,4-Dichlorophenoxyacetic
acid

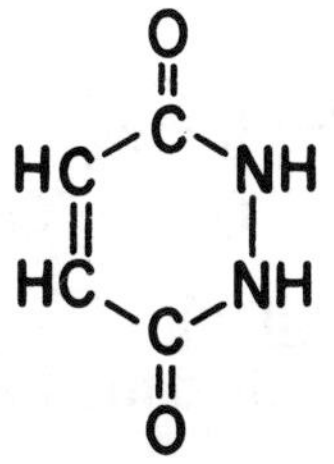

O=C, HC, NH, HC, NH, C=O

MH
Maleic hydrazide

H$_2$C—C(=O)—N(H)—N(CH$_3$)(CH$_3$)
H$_2$C—C(=O)—OH

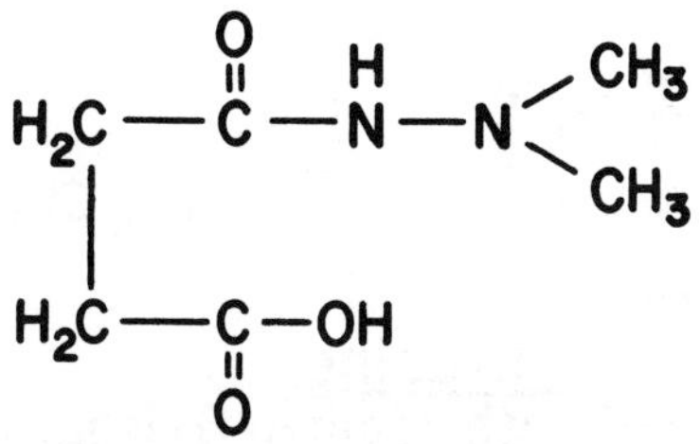

B Nine (B-995)
Succinic acid 2,2-dimethylhydrazide

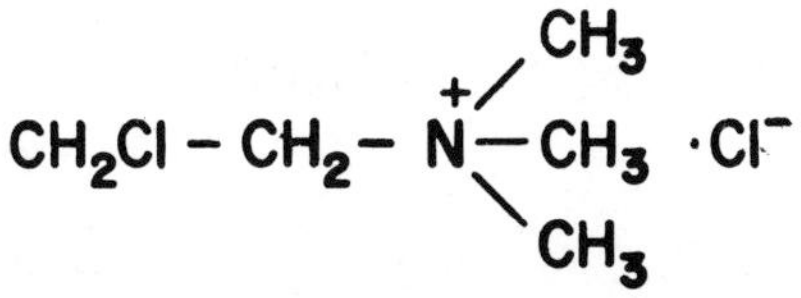

CH$_2$Cl—CH$_2$—N$^+$(CH$_3$)(CH$_3$)(CH$_3$) · Cl$^-$

CCC (Cycocel)
(2-Chloroethyl) trimethylammonium
chloride

Fig. 19-1. Structures of three growth retardants, a growth inhibitor
(MH), and the auxin-type herbicide 2,4-D (redrawn from Frank B. Salisbury
and Cleon Ross, 1969, Wadsworth Publishing Company, Inc., Belmont, California 94002).

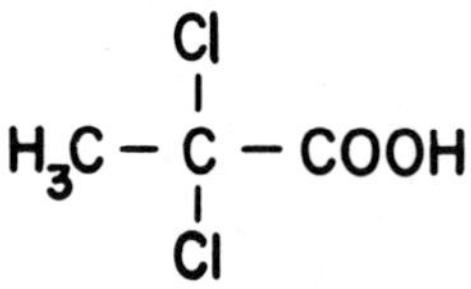

Dalapon
2,2-Dichloropropionic acid

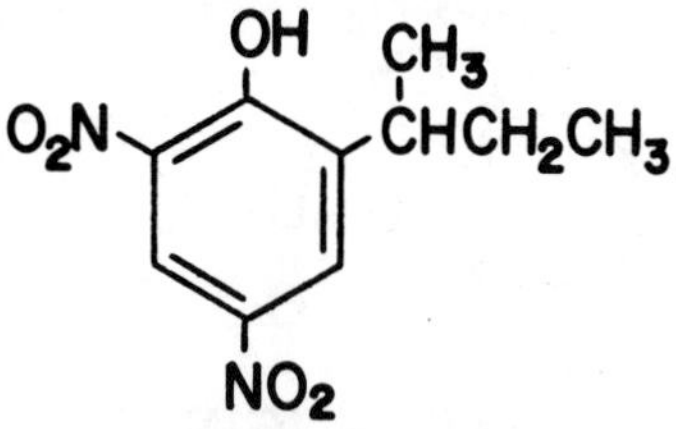

H$_3$C — C — COOH (with Cl above and Cl below the central C)

DNBP
4,6-Dinitro-o-sec-butylphenol

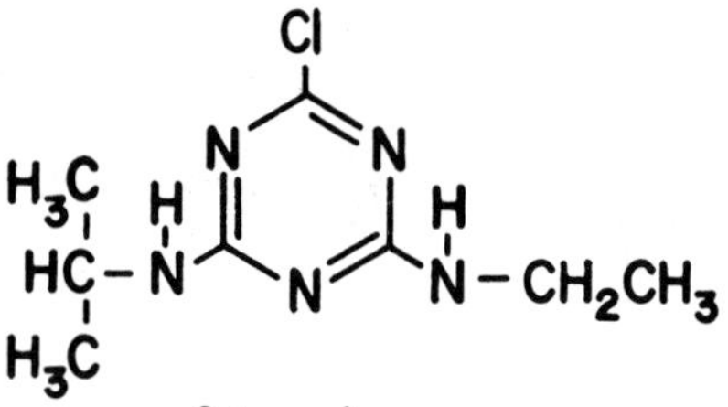

OH, CH$_3$, CHCH$_2$CH$_3$, O$_2$N, NO$_2$

Atrazine
2-Chloro-4-ethylamino-6-isopropyl-
amino-1,3,5-triazine

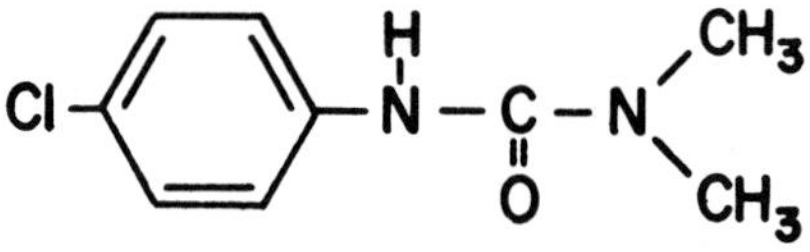

Cl, H$_3$C, HC—N, H$_3$C, N—CH$_2$CH$_3$

Monuron
3(4-Chlorophenyl)1,1-dimethyl urea

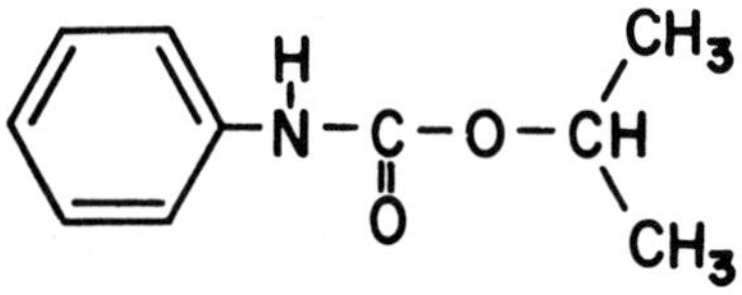

Cl, N—C—N, CH$_3$, CH$_3$, O

IPC
Isopropyl N-phenylcarbamate

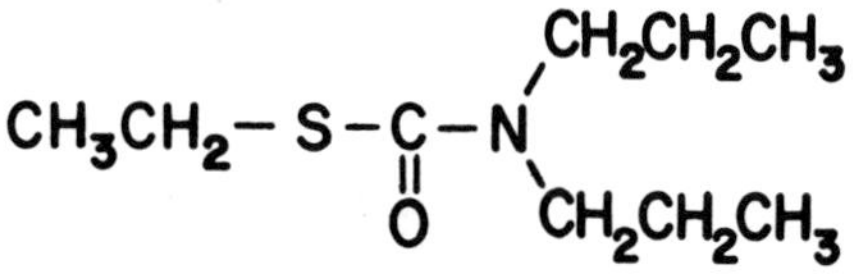

N—C—O—CH, CH$_3$, CH$_3$, O

EPTC
Ethyl N,N-dipropylthiocarbamate

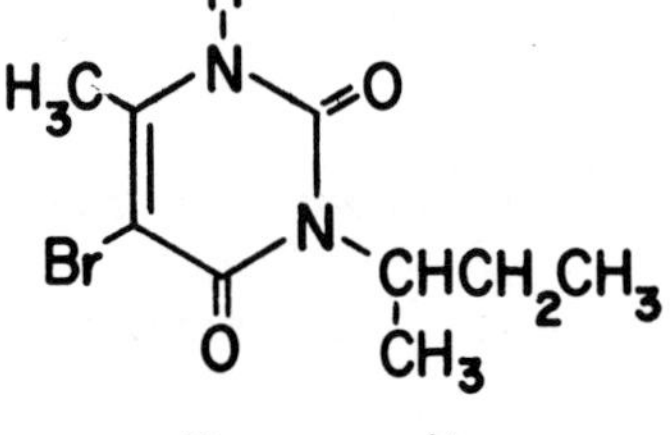

CH$_3$CH$_2$—S—C—N, CH$_2$CH$_2$CH$_3$, CH$_2$CH$_2$CH$_3$, O

Bromacil
5-Bromo-3-sec-butyl-6-methyl
uracil

H$_3$C, H, O, Br, O, CHCH$_2$CH$_3$, CH$_3$

Fig. 19-2. Structures of some non-auxin type herbicides (redrawn from Frank B. Salisbury and Cleon Ross, 1969, Wadsworth Publishing Company, Inc., Belmont, California 94002).

B. Succinic acid 2,2-dimethylhydrazide (B-995, "Alar," N,N-di-
methylaminosuccinamic acid) (Mol. Wt. 160). Spray one pot each of Alaska
peas and barley with an aqueous 10^{-2} M solution (=1,600 mg/1) of B-995[2]
containing 0.05% Tween 20. Also spray one pot of each kind of plant with
0.05% Tween 20 only.

C. (2-Chloroethyl)-trimethylammonium chloride (CCC, "Cycocel") (Mol.
Wt. 158). Apply 500 ml of 10^{-2} M (=1,580 mg/1) CCC in mineral nutrient
solution[3] to the vermiculite rooting medium in one pot each of bush bean,
wheat and barley. Apply 500 ml of nutrient solution only to another pot
of each kind of plant. Supply all plants with CCC solution or plain
mineral nutrient solution as needed for at least one week or until the
end of the experiment.

D. 2,4-Dichlorophenoxyacetic acid (2,4-D) (Mol. Wt. 221). Spray
one pot each of a mixture of Alaska peas plus barley and sunflower plus
wheat with an aqueous 10^{-2} M solution (=2,210 mg/1) of 2,4-D[4] containing
0.05% Tween 20. Spray one pot of each mixture also with 0.05% Tween 20
alone.

E. Amo-1618 [(4-hydroxy-5-isopropyl-2-methylphenyl)-trimethylammonium
chloride 1-piperidine carboxylate] (Mol. Wt. 358). Spray one pot each of
corn and bush bean with an aqueous 10^{-2} M (=3,580 mg/1) solution of
Amo-1618[5] containing 0.05% Tween 20. Also spray one pot of each kind of
plant with 0.05% Tween 20 only.

Check the plants at least once weekly and record observations. Terminate the experiment approximately two weeks after treatment. Measure
the shoot heights of all plants and calculate the mean for each group.
Record any observed effects of the chemicals, such as breaking of apical
dominance, chlorosis, etc. Present the growth data collectively in a
single table.

References

1. Anderson, J. D. and T. C. Moore. 1967. Biosynthesis of (-)-
      kaurene in cell-free extracts of immature pea seeds. Plant
      Physiol. 42: 1527-1534.

---

[2]Dissolve 400 mg B-995 (Alar) in 250 ml distilled water; add 0.1 to
0.2 ml Tween 20 and mix thoroughly.

[3]Dissolve 4.74 g CCC (Cycocel) in 3,000 ml complete mineral nutrient
solution (see Exercise 22 - "Mineral Nutrition of Sunflowers" for description of nutrient solution).

[4]Dissolve 552 mg 2,4-D in 50 ml of 0.1% NaOH and dilute to 250 ml
with distilled water; add 0.1 to 0.2 ml Tween 20 and mix well.

[5]Dissolve 895 mg Amo-1618 in 250 ml distilled water; add 0.1 to 0.2
ml Tween 20 and mix thoroughly.

2. Audus, L. J., Ed. 1964. The Physiology and Biochemistry of Herbicides. Academic Press, New York.

3. Barnes, M. F., E. N. Light, and A. Lang. 1969. The action of plant growth retardants on terpenoid biosynthesis. Planta 88: 172-182.

4. Cathey, H. M. 1964. Physiology of growth retarding chemicals. Ann. Rev. Plant Physiol. 15: 271-302.

5. Coupland, D. and A. J. Peel. 1971. Uptake and incorporation of ^{14}C-labelled maleic hydrazide into the roots of _Salix viminalis_. Physiol. Plantarum 25: 141-144.

6. Crafts, A. S. and W. W. Robbins. 1962. Weed Control. A Textbook and Manual. 3rd Ed. McGraw-Hill Book Company, New York.

7. Dennis, D. T., C. D. Upper, and C. A. West. 1965. An enzymic site of inhibition of gibberellin biosynthesis by Amo 1618 and other plant growth retardants. Plant Physiol. 40: 948-952.

8. Fall, R. R. and C. A. West. 1971. Purification and properties of kaurene synthetase from _Fusarium moniliforme_. J. Biol. Chem. 246: 6913-6928.

9. Galston, A. W. and P. J. Davies. 1970. Control Mechanisms in Plant Development. Prentice-Hall, Englewood Cliffs, New Jersey.

10. Haber, A. H. and J. D. White. 1960. Action of maleic hydrazide on dormancy, cell division, and cell expansion. Plant Physiol. 35: 495-499.

11. Kuraishi, S. and R. M. Muir. 1963. Mode of action of growth retarding chemicals. Plant Physiol. 38: 19-24.

12. Lang, A. 1970. Gibberellins: structure and metabolism. Ann. Rev. Plant Physiol. 21: 537-570.

13. Leopold, A. C. and P. E. Kriedemann. 1975. Plant Growth and Development. 2nd Ed. McGraw-Hill Book Company, New York.

14. Lockhart, J. A. 1962. Kinetic studies of certain anti-gibberellins. Plant Physiol. 37: 759-764.

15. Moore, R. H. 1950. Effects of maleic hydrazide on plants. Science 112: 52-53.

16. Moore, T. C. 1967. Kinetics of growth retardant and hormone interactions in affecting cucumber hypocotyl elongation. Plant Physiol. 42: 677-684.

17. Moore, T. C. 1968. Translocation of the growth retardant N,N-dimethylaminosuccinamic acid-^{14}C (B-995-^{14}C). Botan. Gaz. 129: 280-285.

18. Moore, T. C. 1979. Biochemistry and Physiology of Plant Hormones. Springer-Verlag, New York.

19. Ninnemann, H., J. A. D. Zeevaart, H. Kende, and A. Lang. 1964. The plant growth retardant CCC as inhibitor of gibberellin biosynthesis in Fusarium moniliforme. Planta 61: 229-235.

20. Noodén, L. D. 1969. The mode of action of maleic hydrazide. Inhibition of growth. Physiol. Plantarum 22: 260-270.

21. Noodén, L. D. 1970. Metabolism and binding of ^{14}C-maleic hydrazide. Plant Physiol. 45: 46-52.

22. Phillips, I. D. J. 1971. Introduction to the Biochemistry and Physiology of Plant Growth Hormones. McGraw-Hill Book Company, New York.

23. Reed, D. J., T. C. Moore, and J. D. Anderson. 1965. Plant growth retardant B-995: a possible mode of action. Science 148: 1469-1471.

24. Reid, D. M. and A. Crozier. 1970. CCC-induced increase of gibberellin levels in pea seedlings. Planta 94: 95-106.

25. Riddell, J. A., H. A. Hageman, C. M. J'Anthony, and W. L. Hubbard. 1962. Retardation of plant growth by a new group of chemicals. Science 136: 391.

26. Salisbury, F. B. and C. W. Ross. 1978. Plant Physiology. 2nd Ed. Wadsworth Publishing Company, Belmont, California.

27. Shechter, I. and C. A. West. 1969. Biosynthesis of gibberellins. IV. Biosynthesis of cyclic diterpenes from trans-geranyl-geranyl pyrophosphate. J. Biol. Chem. 244: 3200-3209.

28. Steward, F. C. and A. D. Krikorian. 1971. Plants, Chemicals and Growth. Academic Press, New York.

29. Tolbert, N. E. 1960. (2-Chloroethyl) trimethylammonium chloride and related compounds as plant growth substances. II. Effect on growth of wheat. Plant Physiol. 35: 380-385.

30. Williams, M. W. and E. A. Stahly. 1970. N-malonyl-D-tryptophan in apple fruits treated with succinic acid 2,2-dimethyl-hydrazide. Plant Physiol. 46: 123-125.

31. Woodford, E. K., K. Holly, and C. C. McCready. 1958. Herbicides. Ann. Rev. Plant Physiol. 9: 311-358.

Special Materials and Equipment Required
Per Laboratory Section

*(4) Pots of bush bean (Phaseolus vulgaris) seedlings 9 to 12 days old and rooted in soil

*(4) Pots of corn (Zea mays) seedlings 9 to 12 days old and rooted in soil

*(2) Pots of barley (Hordeum vulgare) seedlings approximately 9 days old and rooted in soil

*(2) Pots of Alaska pea (Pisum sativum) seedlings 9 to 12 days old and rooted in soil

  (2) Pots of wheat (Triticum aestivum) seedlings approximately 9 days old and rooted in vermiculite

  (2) Pots of bush bean seedlings 9 to 12 days old and rooted in vermiculite

  (2) Pots of barley seedlings approximately 9 days old and rooted in vermiculite

*(2) Pots of Alaska pea + barley seedlings (mixed planting) 9 to 12 days old and rooted in soil

*(2) Pots of sunflower (Helianthus annuus) + wheat seedlings (mixed planting) 9 to 12 days old and rooted in soil

*(4 liters) Vermiculite

  (250 ml) MH solution, 4 X 10^{-2} M containing 0.05% Tween 20 (Maleic acid hydrazide available, for example, from Nutritional Biochemicals Corporation, Cleveland, Ohio)

  (250 ml) B-995 (Alar) solution, 10^{-2} M containing 0.05% Tween 20 (Alar available from Naugatuck Chemical Division of U.S. Rubber Company, Naugatuck, Connecticut)

  (250 ml) 2,4-D solution, 10^{-2} M containing 0.05% Tween 20 (2,4-D available, for example, from Mann Research Laboratories, New York, New York

  (3 liters) CCC (Cycocel) solution, 10^{-2} M in nutrient solution (Cycocel available from American Cyanamid Company, Princeton, New Jersey)

  (250 ml) Amo-1618 solution, 10^{-2} M containing 0.05% Tween 20 (Amo-1618 available from Enomoto and Company, Redwood City, California)

  (250 ml) Aqueous 0.05% Tween 20 solution (Tween 20 available from Nutritional Biochemicals Corporation, Cleveland, Ohio)

  (5) Atomizer sprayers (e.g. Windex liquid window cleaner)

  (Stock supply) Complete mineral nutrient solution

Recommendations for Scheduling

It is suggested that this experiment be conducted as a class project in each laboratory section. Seeds should be planted in advance so that seedlings of appropriate age are available for treatment on the day the experiment is to be started.

---

*All plants can be grown in vermiculite and nutrient solution if convenient.

<u>Duration</u>: Approximately 2 weeks (following treatment); 1 hour required in laboratory period when treatments are made, 30 minutes required each succeeding period, and 1 to 1.5 hours required when experiment is terminated.

EFFECTS OF CERTAIN SYNTHETIC PLANT GROWTH REGULATORS ON THE DEVELOPMENT OF SELECTED SPECIES

REPORT

Name _____ Section _____ Date _____

Final measurements of shoot length:

Growth regulator	Conc. (M)	Species	Number of plants measured		Mean final shoot length (cm)	
			Control	Treated	Control	Treated
MH	4×10^{-2}	_____	_____	_____	_____	_____
MH	4×10^{-2}	_____	_____	_____	_____	_____
MH	4×10^{-2}	_____	_____	_____	_____	_____
B-995	10^{-2}	_____	_____	_____	_____	_____
B-995	10^{-2}	_____	_____	_____	_____	_____
B-995	10^{-2}	_____	_____	_____	_____	_____
CCC	10^{-2}	_____	_____	_____	_____	_____
CCC	10^{-2}	_____	_____	_____	_____	_____
CCC	10^{-2}	_____	_____	_____	_____	_____
2,4-D	10^{-2}	_____	_____	_____	_____	_____
2,4-D	10^{-2}	_____	_____	_____	_____	_____
2,4-D	10^{-2}	_____	_____	_____	_____	_____
Amo-1618	10^{-2}	_____	_____	_____	_____	_____
Amo-1618	10^{-2}	_____	_____	_____	_____	_____
Amo-1618	10^{-2}	_____	_____	_____	_____	_____

EFFECTS OF CERTAIN SYNTHETIC PLANT GROWTH REGULATORS
ON THE DEVELOPMENT OF SELECTED SPECIES

REPORT

Name _____ Section _____ Date _____

Observations on plants treated with growth regulators:

A. MH:

B. B-995:

C. CCC:

D. 2,4-D:

E. Amo-1618:

EFFECTS OF CERTAIN SYNTHETIC PLANT GROWTH REGULATORS
ON THE DEVELOPMENT OF SELECTED SPECIES

REPORT

Name _____ Section _____ Date _____

Questions

1. Describe, on the basis of the evidence presently available, the possible modes or mechanisms of action of each of the regulators used.

2. Discuss the differential effects, if any, of B-995, CCC and 2,4-D on the monocot barley and the dicots that you treated.

3. Which of the 5 regulators tested is widely used as a selective herbicide? What is the basis for the selective toxicity of the compound?

EXERCISE 19

EFFECTS OF CERTAIN SYNTHETIC PLANT GROWTH REGULATORS
ON THE DEVELOPMENT OF SELECTED SPECIES

REPORT

Name _____ Section _____ Date _____

4. Which of the 5 compounds tested has been shown to inhibit the bio-synthesis of gibberellin? of auxin?

5. Of what practical use is each of the 5 regulators?

6. Which of the 5 coupounds is classifiable as an "auxin"? Why?

7. For what reason was Tween 20 added to the spray solutions?

Exercise 20

Growth Retardant and Hormone Interactions in Affecting Cucumber Hypocotyl Elongation

Introduction

The growth retardants are a chemically very diverse group of synthetic plant growth-regulating chemicals which are characterized by the common property of inhibiting or "retarding" growth without evoking severe morphological abnormalities. Best known among these compounds are Amo-1618, CCC, Phosphon D and B-995 (Fig. 20-1). Most of the growth retardants are effective on a broad spectrum of angiosperm species. Generally, however, they are more effective on dicots than monocots, and some are effective exclusively on dicots.

Certain of the growth retardants are already being utilized extensively to modify growth of economically and esthetically important plants. The potential for future advances in the practice of altering plant growth and development by the application of chemicals is great. Of keen interest to plant physiologists also is the usefulness of such chemicals in basic research on hormone biosyntheses and hormone-mediated processes. Thus one of the benefits of investigations of the action of growth retardants on

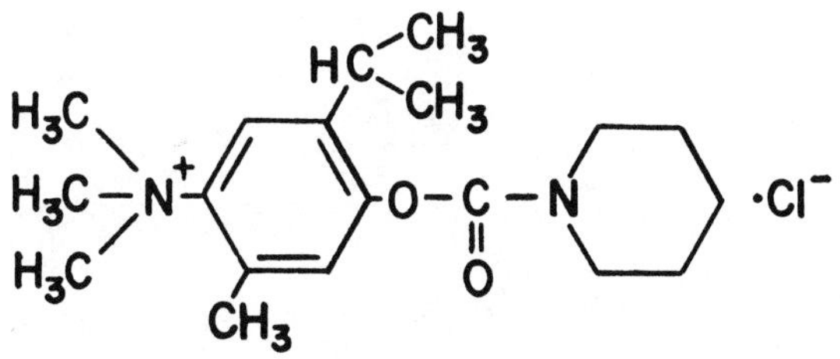

Amo - 1618

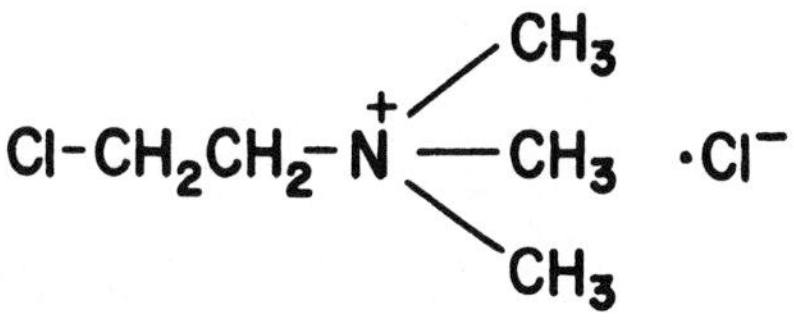

CCC

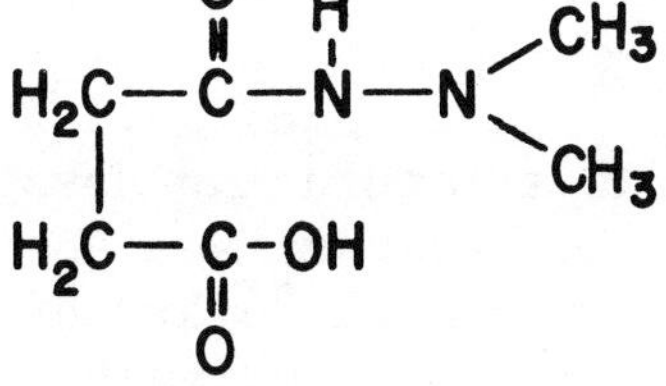

B - 995

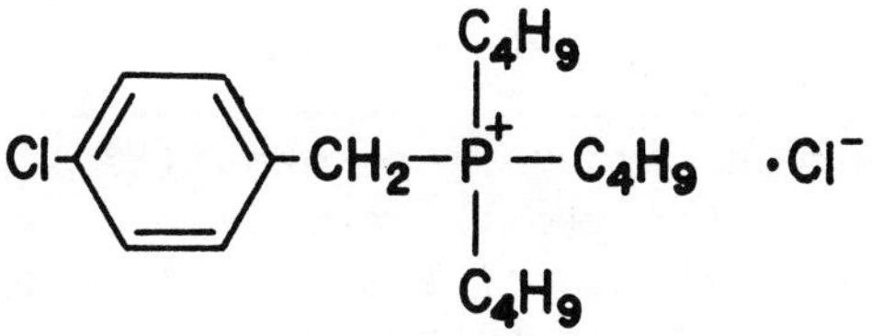

Phosphon - D

Fig. 20-1. Structures of four common growth retardants.

plants is to learn more about the normal hormonal regulation of plant growth and development.

Early in investigations of the physiological action of growth retardants it was hypothesized that the striking effects of the chemicals on stem elongation might be due to their interference with endogenous hormones, specifically the native auxin indoleacetic acid (IAA) and the gibberellins (GAs) (Fig. 20-2), in the plant. Conceivably, such interference might be with transport, biosynthesis, metabolic degradation or mechanism of action of endogenous auxin or GA or both. Some evidence has been obtained that certain growth retardants do interfere with the biosynthesis or subsequent metabolism of endogenous hormones. In fact, the most convincing evidence yet obtained regarding the modes of action of Amo-1618, CCC and Phosphon D is inhibition of GA biosynthesis, although this may not be the exclusive mode of action of any of the compounds.

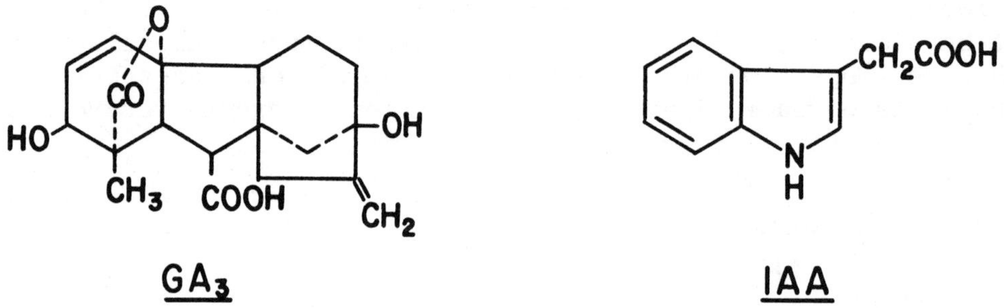

Fig. 20-2. Structures of gibberellin A₃ and indoleacetic acid.

If, in fact, a particular growth retardant does interact directly with a specific endogenous hormone, by inhibiting synthesis, for example, it should be possible to demonstrate direct ("competitive") interaction in vivo in an appropriate whole plant. An appropriate whole plant in this case would be one which exhibits positive stem elongation responses to exogenous IAA and GA independently. The basic experimental design would be to use seedling plants which are displaying a constant (or nearly constant) rate of stem elongation, and to treat the plants with varying dosages of hormone, from zero to saturation, in the presence and absence of a standard dosage of the growth retardant (Fig. 20-3). If at a saturating dosage of the hormone the growth rate is the same in both the presence and absence of the growth retardant; that is, if the effect of the growth retardant is nullified by a saturating dosage of hormone, it may be tentatively concluded that a direct interaction does occur (Fig. 20-3A). If, however, the inhibition by the growth retardant is the same in plants saturated with exogenous hormone as in those with only endogenous hormone; that is, if the effect of the growth retardant is not nullified or overcome at saturating dosages

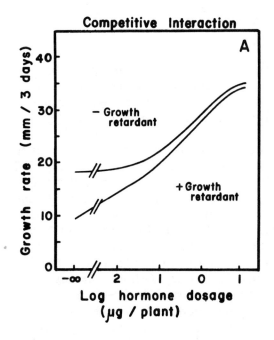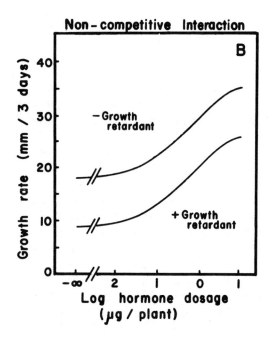

Fig. 20-3. Illustrations of the types of results obtained
in kinetic analyses of growth hormone-growth retardant interactions in an appropriate test system. A, Direct or "competitive" interaction. B, "Non-competitive" interaction. Results
obtained with particular hormones and growth retardants which
do not interact directly do not fit parallel curves as depicted
in B, but show divergence at increasing dosages of the hormone.

of hormone, it may be concluded that there is no direct interaction of the
hormone and the growth retardant (Fig. 20-3B). It must be emphasized that
such a kinetic analysis of growth retardant-hormone interactions reveals
no direct information on the specific nature of, or biochemical basis for,
an observed growth retardant-hormone interaction. This information can be
gained only through further, more detailed types of investigations.

Whole seedlings of but a very few species of plants can be used in
the type of kinetic analysis described previously, because whole seedlings
of most species do not exhibit positive stem elongation responses to
exogenous IAA and some do not exhibit responses to exogenous GA. Cucumber
seedlings of the National Pickling variety are a very useful exception,
therefore, since hypocotyl elongation in these seedlings is promoted by
several auxins as well as several gibberellins.

The purpose of this experiment (actually four related experiments) is
to investigate possible interactions of two growth retardants, Amo-1618 and
CCC, with IAA and GA$_3$ in hypocotyl elongation in intact cucumber seedlings.

Materials and Methods[1]

This exercise consists of 4 experiments, two to test the possible interaction of Amo-1618 with GA$_3$ and with IAA, and two to test the possible interaction of CCC with GA$_3$ and IAA. Each team will perform either one or two experiments, according to the directions of the instructor.

For all 4 experiments the same procedures should be followed in growing and handling the plants. For each experiment, soak approximately 250 cucumber seeds for 2 to 3 hours in distilled water. Plant approximately 25 seeds in each of 10 plastic pots (approximately 6-inch diameter at top and 4 1/2-inch depth) filled with vermiculite. Each pot should be equipped with a saucer or tray to catch excess solution that drains out. Add about 500 ml of complete mineral nutrient solution[2] to each container; after saturation of the vermiculite allow excess solution to drain out. Grow the plants in a growth chamber or other facility in a 16-hour photoperiod at approximately 30° C and a light intensity of about 600 ft-c and an 8-hour dark period at approximately 27° C. When additional nutrient solution is needed prior to the time of treatment, add an equal volume to each pot. Routinely, it should be necessary to add about 100 ml of solution at 2-day intervals.

Five to 6 days after planting (or when the hypocotyls are 2.5 to 3 cm long and the cotyledons are about fully expanded), thin the seedlings to leave 12 to 15 highly uniform seedlings in each pot. Mark the hypocotyl of each seedling gently with India ink at the cotyledonary node and 2.0 cm below the cotyledonary node. This section is designated the hypocotyl unit. Treat the plants promptly, as described below for each experiment, and return the seedlings to the growth chamber for an additional 3 days. At the end of 3 days, measure the length of the hypocotyl units.

Experiment 1. Interaction between Amo-1618 and GA$_3$. Divide the 10 pots of seedlings into 2 series of 5 pots each. To the shoot tip of each seedling in one series, apply 10 μl (microliters) of Amo-1618 solution,[3] using a micrometer buret of approximately 0.2-ml capacity. To the shoot tip of each seedling in the other series apply 10 μl of control solution (aqueous 0.05% Tween 20). Later in the same laboratory period, apply to the shoot tips of the plants in one pot of each of the 2 series (with and

---

[1]Adopted from: Katsumi, M., B. O. Phinney, and W. K. Purves. 1965. The roles of gibberellin and auxin in cucumber hypocotyl growth. Physiol. Plantarum 18: 462–473; and Moore, T. C. 1967. Kinetics of growth retardant and hormone interactions in affecting cucumber hypocotyl elongation. Plant Physiol. 42: 677–684.

[2]See Exercise 22 – "Mineral Nutrition of Sunflowers" for composition of nutrient solution.

[3]Prepare 10 ml of Amo-1618 solution at a concentration of 1 mg/ml by dissolving 10 mg of growth retardant in 10 ml of distilled water containing 0.05% Tween 20. Ten microliters (10 μl) then contains 10 μg of Amo-1618.

without Amo-1618 treatment) 10 µl each of one of the following dosages of GA_3:[4] 0, 0.01, 0.1, 1.0 and 10 µg. Label all 10 groups clearly according to the specific treatments they received.

Experiment 2. Interaction between Amo-1618 and IAA. Follow the same procedure exactly as described for Experiment 1, except substitute IAA[5] for GA_3. Precisely the same procedure described for the preparation of GA_3 solutions should be followed in preparing IAA solutions.

Experiment 3. Interaction between CCC and GA_3. Divide the 10 pots into 2 series of 5 pots each. To the vermiculite in each pot of one series, add 250 ml of complete mineral nutrient solution containing 10^{-3} M CCC.[6] Add 250 ml of plain nutrient solution to each pot of the other series. No further addition of solution is likely to be needed before the experiment is terminated, but, if so, add an equal volume of the appropriate solution to each pot. Later in the same laboratory period, treat the plants in one pot of each of the 2 series with 10 µl containing 0, 0.01, 0.1, 1.0 or 10 µg of GA_3, using a micrometer buret and applying solution directly to the shoot tip.

Experiment 4. Interaction between CCC and IAA: Follow the same procedures described for Experiment 3, except substitute IAA for GA_3.

Each student record the summary data and plot graphs for all 4 experiments. Each graph should contain 2 curves, one for hormone only and one for hormone plus growth retardant (see examples in Fig. 20-3).

References

1. Anderson, J. D. and T. C. Moore. 1967. Biosynthesis of (-)-kaurene in cell-free extracts of immature pea seeds. Plant Physiol. 42: 1527-1534.

---

[4]Prepare a GA_3 stock solution by dissolving 10 mg of GA_3 in 0.5 to 1.0 ml of 95% ethanol and diluting with distilled water containing 0.05% Tween 20 to a final volume of 10 ml. Then make serial dilutions of the stock solution with aqueous 0.05% Tween 20 to give 9 or 10 ml of solution at concentrations of 0.001, 0.01 and 0.1 mg GA_3/ml. Ten µl of the solutions, including the stock solution, then contain 0.01, 0.1, 1.0 and 10 µg GA_3, respectively. Control plants are treated with aqueous 0.05% Tween 20.

[5]Prepare an IAA stock solution by dissolving 10 mg of IAA in 0.5 to 1.0 ml of 95% ethanol and diluting with distilled water containing 0.05% Tween 20 to a final volume of 10 ml. Then make serial dilutions of the stock solution with aqueous 0.05% Tween 20 to give 9 or 10 ml of solution at concentrations of 0.001, 0.01 and 0.1 mg IAA/ml. Ten µl of the solutions, including the stock solution, then contain 0.01, 0.1, 1.0 and 10 µg IAA, respectively. Control plants are treated with aqueous 0.05% Tween 20.

[6]Prepare 2.5 liters of CCC solution by dissolving 395 mg of CCC (Mol. wt. = 158) directly in 2.5 liters of complete mineral nutrient solution.

2. Barnes, M. F., E. N. Light, and A. Lang. 1969. The action of plant growth retardants on terpenoid biosynthesis. Planta 88: 172-182.

3. Cathey, H. M. 1964. Physiology of growth retarding chemicals. Ann. Rev. Plant Physiol. 15: 271-302.

4. Cleland, R. 1965. Evidence on the site of action of growth retardants. Plant Cell Physiol. 6: 7-15.

5. Cleland, R. E. 1969. The gibberellins. Pp. 49-81 in: M. B. Wilkins, Ed. The Physiology of Plant Growth and Development. McGraw-Hill Book Company, New York.

6. Dennis, D. T., C. D. Upper, and C. A. West. 1965. An enzymic site of inhibition of gibberellin biosynthesis by Amo-1618 and other plant growth retardants. Plant Physiol. 40: 948-952.

7. Drury, R. E. 1969. Interaction of plant hormones. Science 164: 564-565.

8. Fall, R. R. and C. A. West. 1971. Purification and properties of kaurene synthetase from Fusarium moniliforme. J. Biol. Chem. 246: 6913-6928.

9. Galston, A. W. and P. J. Davies. 1970. Control Mechanisms in Plant Development. Prentice-Hall, Englewood Cliffs, New Jersey.

10. Halevy, A. H. and H. M. Cathey. 1960a. Effects of structure and concentration of gibberellins on the growth of cucumber seedlings. Botan. Gaz. 122: 63-67.

11. Halevy, A. H. and H. M. Cathey. 1960b. Effect of structure and concentration of some quaternary ammonium compounds on growth of cucumber seedlings. Botan. Gaz. 122: 151-154.

12. Harada, H. and A. Lang. 1965. Effect of some (2-chloroethyl) trimethylammonium chloride analogs and other growth retardants on gibberellin biosynthesis in Fusarium moniliforme. Plant Physiol. 40: 176-183.

13. Katsumi, M., B. O. Phinney, and W. K. Purves. 1965. The roles of gibberellin and auxin in cucumber hypocotyl growth. Physiol. Plantarum 18: 462-473.

14. Katsumi, M., W. K. Purves, B. O. Phinney, and J. Kato. 1965. The role of the cotyledons in gibberellin- and auxin-induced growth of the cucumber hypocotyl. Physiol. Plantarum 18: 550-556.

15. Kende, H., H. Ninnemann, and A. Lang. 1963. Inhibition of gibberellic acid biosynthesis in Fusarium moniliforme by Amo-1618 and CCC. Naturwissenschaften 50: 599-600.

16. Kuraishi, S. and R. M. Muir. 1963. Mode of action of growth
    retarding chemicals. Plant Physiol. 38: 19-24.

17. Lang, A. 1970. Gibberellins: structure and metabolism. Ann.
    Rev. Plant Physiol. 21: 537-570.

18. Leopold, A. C. and P. E. Kriedemann. 1975. Plant Growth and
    Development. 2nd Ed. McGraw-Hill Book Company, New York.

19. Lockhart, J. A. 1962. Kinetic studies of certain anti-gibber-
    ellins. Plant Physiol. 37: 759-764.

20. Lockhart, J. A. 1965. The analysis of interactions of physical
    and chemical factors on plant growth. Ann. Rev. Plant
    Physiol. 16: 37-52.

21. Moore, T. C. 1967. Kinetics of growth retardant and hormone
    interactions in affecting cucumber hypocotyl elongation.
    Plant Physiol. 42: 677-684.

22. Moore, T. C. 1979. Biochemistry and Physiology of Plant
    Hormones. Springer-Verlag, New York.

23. Norris, R. F. 1966. Effect of (2-chloroethyl) trimethylammonium
    chloride on the level of endogenous indole compounds in
    wheat seedlings. Can. J. Botany 44: 675-684.

24. Phillips, I. D. J. 1971. Introduction to the Biochemistry and
    Physiology of Plant Growth Hormones. McGraw-Hill Book
    Company, New York.

25. Reid, D. M. and A. Crozier. 1970. CCC-induced increase of
    gibberellin levels in pea seedlings. Planta 94: 95-106.

26. Sachs, R. M. and M. A. Wohlers. 1964. Inhibition of cell
    proliferation and expansion in vitro by three stem growth
    retardants. Am. J. Botany 51: 44-48.

27. Shechter, I. and C. A. West. 1969. Biosynthesis of gibberellins.
    IV. Biosynthesis of cyclic diterpenes from trans-geranyl-
    geranyl pyrophosphate. J. Biol. Chem. 244: 3200-3209.

Special Materials and Equipment Required
Per Team of 3-6 Students

(250 or 500) Cucumber (Cucumis sativus) seeds of the National Pickling
    variety (available from W. Atlee Burpee Company, Philadelphia,
    Pennsylvania, Clinton, Iowa or Riverside, California)
(10 or 20) Plastic pots, approximately 6 inches in diameter, and drain
    saucers
(10 or 20 liters) Vermiculite
(Stock supply) Complete mineral nutrient solution
(1) Growth chamber or other controlled-environment facility
(1) India ink pen or other marking device

(1 to 3) Micrometer burets (e.g. 0.2 ml capacity micrometer buret available from Roger Gilmont Instruments, Incorporated, Vineland, New Jersey 08360)

(10 or 20 ml) Amo-1618 solution, prepared as described in Materials and Methods

(10 or 20 ml) Each GA solution, prepared as described in Materials and Methods

(10 or 20 ml) Each IAA solution, prepared as described in Materials and Methods

(10 or 20 ml) Aqueous 0.05% Tween 20 solution

(2.5 or 5 liters) CCC solution, prepared with complete mineral nutrient solution, as described in Materials and Methods

Recommendations for Scheduling

Direct students to work in teams of 3 to 6 members, each team performing one or two complete experiments. Solutions of growth regulators can either be prepared in advance of the laboratory period in which they are required or be prepared by the students.

Duration: Approximately 8 days; 1 hour required during period when seeds are planted, 1.5 to 2 hours required at time of treatment, and 1 to 1.5 hours required when experiment is terminated.

EXERCISE 20

GROWTH RETARDANT AND HORMONE INTERACTIONS IN AFFECTING CUCUMBER HYPOCOTYL ELONGATION

REPORT

Name _____ Section _____ Date _____

Final measurements of hypocotyl units:

Experiment 1 - Interaction between Amo-1618 and GA:

GA treatment (µg/plant)	Average final length (mm)		Average increment (mm)	
	−Amo−1618	+Amo−1618	−Amo−1618	+Amo−1618
0	_____	_____	_____	_____
0.01	_____	_____	_____	_____
0.1	_____	_____	_____	_____
1.0	_____	_____	_____	_____
10	_____	_____	_____	_____

Experiment 2 - Interaction between Amo-1618 and IAA:

IAA treatment (µg/plant)	Average final length (mm)		Average increment (mm)	
	−Amo−1618	+Amo−1618	−Amo−1618	+Amo−1618
0	_____	_____	_____	_____
0.01	_____	_____	_____	_____
0.1	_____	_____	_____	_____
1.0	_____	_____	_____	_____
10	_____	_____	_____	_____

EXERCISE 20

GROWTH RETARDANT AND HORMONE INTERACTIONS IN AFFECTING CUCUMBER HYPOCOTYL ELONGATION

REPORT

Name _____ Section _____ Date _____

Experiment 3 - Interaction between CCC and GA:

GA Treatment (µg/plant)	Average final length (mm)		Average increment (mm)	
	- CCC	+ CCC	- CCC	+ CCC
0				
0.01				
0.1				
1.0				
10				

Experiment 4 - Interaction between CCC and IAA:

IAA treatment (µg/plant)	Average final length (mm)		Average increment (mm)	
	- CCC	+ CCC	- CCC	+ CCC
0				
0.01				
0.1				
1.0				
10				

238

EXERCISE 20

GROWTH RETARDANT AND HORMONE INTERACTIONS IN AFFECTING
CUCUMBER HYPOCOTYL ELONGATION

REPORT

Name _____   Section _____   Date _____

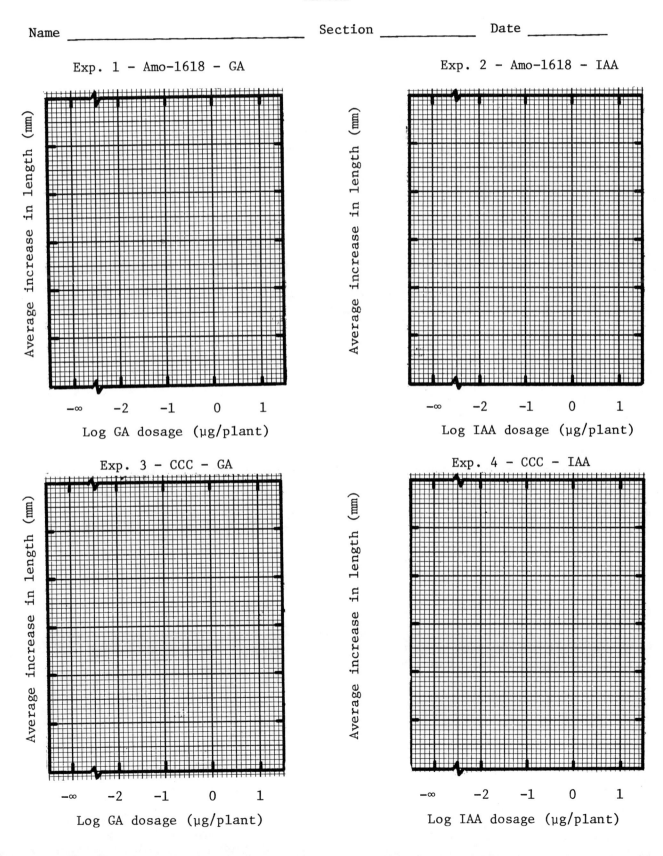

Exp. 1 – Amo–1618 – GA

Average increase in length (mm)

−∞   −2   −1   0   1

Log GA dosage (µg/plant)

Exp. 2 – Amo–1618 – IAA

Average increase in length (mm)

−∞   −2   −1   0   1

Log IAA dosage (µg/plant)

Exp. 3 – CCC – GA

Average increase in length (mm)

−∞   −2   −1   0   1

Log GA dosage (µg/plant)

Exp. 4 – CCC – IAA

Average increase in length (mm)

−∞   −2   −1   0   1

Log IAA dosage (µg/plant)

EXERCISE 20

GROWTH RETARDANT AND HORMONE INTERACTIONS IN AFFECTING CUCUMBER HYPOCOTYL ELONGATION

REPORT

Name _____ Section _____ Date _____

Questions

1. From the results of Experiments 1 and 2, what tentative conclusions are warranted regarding possible direct interactions between Amo-1618 and GA and between Amo-1618 and IAA?

2. What tentative conclusions are in order regarding possible direct interactions between CCC and GA and between CCC and IAA, based on the results of Experiments 3 and 4?

3. In what specific ways have CCC and Amo-1618 been reported to interfere with endogenous GA?

4. Based on the collective results of all 4 experiments, does it appear that the mode of action of CCC is exclusively interference with endogenous GA? Explain.

Exercise 20

Growth Retardant and Hormone Interactions in Affecting Cucumber Hypocotyl Elongation

REPORT

Name _____ Section _____ Date _____

5. In what way is the rationale of these experiments, as described in the Introduction, analogous to investigations of enzyme kinetics?

6. Briefly describe the experimental approach you would use in efforts to ascertain the specific metabolic sites and mechanisms of action of Amo-1618 and CCC.

EXERCISE 21

PHOTOPERIODIC CONTROL OF FLOWERING IN COCKLEBUR
(XANTHIUM STRUMARIUM)

Introduction

Photoperiodism[1] may be defined simply as a response to the relative lengths of day and night, or more specifically, as a response to the duration and timing of light and dark periods. This phenomenon is an integral part of the overall adaptation of many kinds of organisms--plant and animal--to their natural environments. Among the kinds of organisms which exhibit photoperiodism are at least some species among non-vascular and vascular cryptogams, gymnosperms, angiosperms, insects, fishes, reptiles, birds and mammals.

In all but equatorial areas, of course, daylength changes at a constant rate with the change of the seasons. Photoperiod is therefore the "least variable variable in nature," as has often been noted. When this fact is appreciated, it does not seem at all surprising that photoperiodism has been of such prominent significance in the evolution of both plant and animal kingdoms.

Many processes and developmental events in both the vegetative and flowering stages of the life cycles of angiosperms are either controlled or greatly influenced by the duration and timing of daily light and dark periods. Among the many known examples of aspects of vegetative development which are affected by daylength, in at least some species, are: (a) onset and breaking of bud dormancy in woody perennials; (b) leaf abscission in deciduous trees and shrubs; (c) bulb and tuber formation in herbaceous species; (d) seed germination; (e) cessation of cambial activity; and (f) development of frost resistance. Such purely vegetative responses frequently are termed vegetative photoperiodism.

More familiar to most students of biology, perhaps, is the effect of daylength on the flowering behavior of many species of angiosperms. It was primarily this phenomenon of reproductive photoperiodism in angiosperms which was described in the classical work of W. W. Garner and H. A. Allard in 1920. Garner and Allard discovered that angiosperms could be classified in three groups, according to the ways by which flowering is affected by daylength. These groups are the short-day plants, long-day plants and day-neutral plants. The short-day plants were said to be those which flower in response to daylengths shorter than some maximum. Long-day plants were identified as those that flower in response to daylengths longer than some critical duration. Day-neutral plants were regarded as those in which flowering is not specifically affected by daylength.

Actually, and as might have been expected, the diversity of response types is much greater than is reflected in the simple classification

---

[1]Animal scientists frequently use the equivalent term "photoperiodicity."

initially devised by Garner and Allard. For example, a few species are known which flower in response to a period of short days followed by a period of long days (short-long-day plants), and a few others flower in response to long days followed by short days (long-short-day plants). Some flower only in response to intermediate daylength. Still a very few others are known to flower in response to either relatively long or relatively short--but not intermediate--daylength.

Complicating the situation further, there are in the categories of both short-day plants and long-day plants species and varieties which exhibit more or less obligatory or qualitative responses and others which exhibit only facultative or quantitative responses. An obligate short-day plant typically has quite exacting requirements for photoperiodic floral induction and will not flower in their absence, whereas a facultative short-day plant does not exhibit an all-or-none response but flowers more readily or more vigorously if it experiences relatively short day-length.

In the preceding discussion emphasis has been placed on daylength. This is somewhat misleading, however, in that photoperiodism is, as stated earlier, a response to the duration and timing of light and dark periods. The single most important controlling factor is the length of the uninterrupted dark period. For all photoperiodic responses occurring in natural 24-hour light-dark cycles, it is possible to speak of both critical daylength and critical nightlength. Critical daylength, in the case of a long-day plant, is the minimum daylength required for flowering; in the case of a short-day plant, it is the maximum daylength on which flowering can occur. The critical nightlength for a short-day plant is the minimum period of darkness required for flowering; for a long-day plant it is the maximum period of darkness on which flowering can occur. Thus for short-day plants the daylength must be shorter than the critical daylength and the dark period longer than the critical nightlength if flowering is to occur. For the long-day plant, on the contrary, the day-length must be longer than the critical daylength and the dark period shorter than the critical nightlength for flowering to occur. The distinction between short-day plants and long-day plants has nothing to do with the absolute values of the critical daylength; the distinction is whether flowering is promoted by daylengths shorter or longer than the critical daylength. For example, the critical daylength for Xanthium strumarium, an absolute short-day plant, is about 15.5 hours, and for Hyoscyamus niger, an obligate long-day plant, about 11 hours. Both species flower well in response to 12- to 14-hour photoperiods in 24-hour light-dark cycles.

A general summary description of the requirements for photoperiodic induction of flowering in obligate short-day plants and long-day plants can be stated as follows:

> Short-day plants - Require exposure to a certain minimum number of photoinductive cycles (which must be consecutive for maximum effectiveness) in each of which an uninterrupted dark period equals or exceeds a critical duration, after attaining "ripeness to flower."

Long-day plants - Require exposure to a certain minimum
number of cycles (not necessarily consecutive) in each of
which the photoperiod equals or exceeds some critical
duration, after attaining "ripeness to flower;" do not
require a dark period.

Examples of obligate short-day plants are cocklebur (Xanthium
strumarium), Japanese morning glory (Pharbitis nil), kalanchoë (Kalanchoë
blossfeldiana), poinsettia (Euphorbia pulcherrima), Maryland Mammoth
tobacco (Nicotiana tabacum) and the common weed Chenopodium rubrum. Among
known obligate long-day plants are black henbane (Hyoscyamus niger), Darnel
ryegrass (Lolium temulentum), sedum (Sedum spectabile), spinach (Spinacia
oleracea), and some varieties of barley (Hordeum vulgare). Tomato
(Lycopersicum esculentum) and corn (Zea mays) are familiar day-neutral
plants.

The age or amount of development which a plant must attain before it
is "ripe to flower" varies markedly with species. Some plants, such as
Pharbitis nil and Chenopodium rubrum can be induced to flower in the seedling stage. Others, such as Xanthium strumarium and Hyoscyamus niger
must be a few weeks of age, and some trees must be several years old
before they can be induced to flower.

Species also differ markedly with respect to the number of photoinductive cycles required for flowering. Some very sensitive species,
such as Xanthium strumarium, Lolium temulentum and Pharbitis nil, will
respond to a single photoinductive cycle. Others require several cycles,
and for maximum effectiveness, in the case of short-day plants, the
cycles must be consecutive to be maximally effective. In those species
which respond to a single photoinductive cycle, as in other species, the
response (rate and vigor) increases as the number of cycles increases, up
to some maximum number.

The leaves (or modified leaves such as cotyledons) are the organs of
the plant that perceive the photoperiodic stimulus. There is abundant
evidence, although no conclusive proof as yet, that in at least some
plants (e.g. Xanthium strumarium) a specific flowering hormone ("florigen"
or "anthesin") is produced in the photoinduced leaf and is translocated
via the phloem to the bud, where it evokes differentiation of a vegetative
meristem into a floral (or inflorescence) primordium.

Current evidence indicates that circadian rhythms play a fundamental
role in governing the flowering response in angiosperms generally. Photoperiodic control appears to be superimposed upon this basic timing mechanism in the photoperiodically sensitive species. Thus the nature of the
control of flowering appears to reside in a complex interaction between
the phytochrome system and circadian rhythms. Phytochrome has been
definitely established as the photoreceptor pigment for photoperiodic
flowering responses, as well as for all other photoperiodic responses.

The purpose of this experiment is to investigate the photoperiodic
control of floral initiation in the short-day plant Xanthium strumarium.

Materials and Methods[2]

Plant two flats of sand or vermiculite with cocklebur fruits (burs) in the greenhouse. By means of incandescent or fluorescent lamps connected to an automatic timer, provide a 20-hour photoperiod (4 AM to 12 PM) and a 4-hour dark period for the plants.

When the plants are two to three weeks of age, transplant 72 of them individually to 4-inch plastic or clay pots filled with soil. Continue to grow the plants under a 20-hour photoperiod.

Approximately six weeks after planting, or any time after the first two leaves of the plants are half-expanded, divide the 72 plants into 6 groups of 12 plants each. Treat one group in each of the following ways:

1. No photoinductive cycles – controls.
2. 1 photoinductive cycle.
3. 2 consecutive photoinductive cycles.
4. 4 consecutive photoinductive cycles.
5. 8 consecutive photoinductive cycles.
6. 1 photoinductive cycle, with dark period interrupted midway by a flash of light.[3]

Conduct the photoinductive treatment in the following manner: Leave all plants on a greenhouse bench under the light except for the daily dark period during actual photoinduction. A light-tight cabinet will be available in the greenhouse for the dark treatment. A photoperiodic cycle consisting of 8 hours of light and 16 hours of darkness will be employed as the photoinductive cycle. To provide plants with a single photoinductive cycle, transfer them to the light-tight cabinet at 5 PM on a given day, then remove and transfer them back to the greenhouse bench at 9 AM on the following day. Thus, a schedule for the treatment to be used in this experiment may be tabulated as follows:

Day 1 – Transfer groups 2, 3, 4 and 5 to cabinet at 5 PM; remove at 9 AM the next day.
Day 2 – Transfer groups 3, 4 and 5 to cabinet at 5 PM; remove at 9 AM the next day.

---

[2]An analogous alternative experiment can be done with Japanese morning glory (Pharbitis nil), strain Violet, another photoperiodically very sensitive short-day plant. For general procedures consult: Takimoto, A. and K. C. Hammer. 1964. Effect of temperature and preconditioning on photoperiodic response of Pharbitis nil. Plant Physiol. 39: 1024-1030. Seeds can be purchased from biological supply houses, e.g. Carolina Biological Supply Company, Burlington, North Carolina 27215.

[3]An alternative procedure for interrupting the dark period is merely to open the dark cabinet at midnight and expose a camera flash bulb directly on the plants at close range.

Day 3 - Transfer groups 4 and 5 to cabinet at 5 PM; remove at 9 AM
     the next day.
Day 4 - Repeat operation of Day 3.
Day 5 - Transfer group 5 to cabinet at 5 PM; remove at 9 AM the next
     day.
Day 6 - Repeat operation of Day 5.
Day 7 - Repeat operation of Day 6.
Day 8 - Repeat operation of Day 7.
Day 9 - Transfer group 6 to cabinet at 6 PM. At midnight, open the
     cabinet, transfer the plants to the greenhouse bench, turn
     on the lights for a period of 15 minutes to 1 hour, then
     return the plants to the dark cabinet; remove plants at 6 AM
     the following morning.[3]

Beginning one week after the termination of the photoinduction period,
examine the shoot tips of the plants for developing terminal inflorescences.
Check the plants at least once a week and record your observations each
time. Approximately four weeks after the start of the photoinduction
period, terminate the experiment. Record the number (and percent) of plants
flowering in each group. Present these data in a table. Then remove all
inflorescences from the plants in each group, weigh and determine the
average fresh weight of floral parts per flowering plant in each group.
Plot a graph showing average fresh weight of floral parts per flowering
plant versus number of photoinductive cycles.

If time permits and plants are available, other interesting experimental manipulations will be made. For example, groups of plants will be
variously debladed (leaf blades removed) just prior to being exposed to a
single inductive dark period. Such an experiment will reveal that a completely debladed plant will not respond to the dark period by flowering,
whereas the presence of as little as half of one half-grown leaf blade will
be enough to permit the plant to respond to the single dark period by flowering. Hence, it will be shown that the leaf blade is the locus of photoperiodic induction, that is, the organ which perceives the photoperiodic
stimulus.

Additionally, evidence that floral initiation in the cocklebur is
mediated by a transmissible stimulus, or flowering hormone, will be obtained
by grafting a leaf from a photoinduced cocklebur plant to a plant kept under
non-inductive light cycles.

References

1. Borthwick, H. A. and R. J. Downs. 1964. Roles of active phyto-
     chrome in control of flowering of Xanthium pennsylvanicum.
     Botan. Gaz. 125: 227-231.

2. Cathey, H. M. and H. A. Borthwick. 1964. Significance of dark
     reversion of phytochrome in flowering of Chrysanthemum
     morifolium. Botan. Gaz. 125: 232-236.

3. Chailakhyan, M. K. 1968. Internal factors of plant flowering.
    Ann. Rev. Plant Physiol. 19: 1-36.

4. Cumming, B. G. 1959. Extreme sensitivity of germination and
    photoperiodic reaction in the genus Chenopodium (Tourn.) L.
    Nature 184: 1044-1045.

5. Cumming, B. G. 1963. Evidence of a requirement for phytochrome-
    P_{fr} in the floral initiation of Chenopodium rubrum. Can. J.
    Botany 41: 901-926.

6. Cumming, B. G. and E. Wagner. 1968. Rhythmic processes in
    plants. Ann. Rev. Plant Physiol. 19: 381-416.

7. Evans, L. T. 1958. Lolium temulentum L., a long-day plant
    requiring only one inductive photocycle. Nature 182: 197-
    198.

8. Evans, L. T., Ed. 1969. The Induction of Flowering. Some Case
    Histories. Cornell University Press, Ithaca, New York.

9. Fredericq, F. 1964. Conditions determing effects of far-red
    and red irradiations on flowering response of Pharbitis nil.
    Plant Physiol. 39: 812-816.

10. Friend, D. J. C. and V. A. Helson. 1966. Brassica campestris
    L.: floral induction by one long day. Science 153: 1115-
    1116.

11. Galston, A. W. and P. J. Davies. 1970. Control Mechanisms in
    Plant Development. Prentice-Hall, Englewood Cliffs, New
    Jersey.

12. Garner, W. W. and H. A. Allard. 1920. Effect of the relative
    length of day and night and other factors of the environment
    on growth and reproduction in plants. J. Agr. Res. 18:
    553-606.

13. Garner, W. W. and H. A. Allard. 1923. Further studies in photo-
    periodism, the response of the plant to the relative length
    of day and night. J. Agr. Res. 23: 871-920.

14. Halaban, R. 1968. The flowering response of Coleus in relation
    to photoperiod and the circadian rhythm of leaf movement.
    Plant Physiol. 43: 1894-1898.

15. Hillman, W. S. 1962. The Physiology of Flowering. Holt, Rinehart
    and Winston, New York.

16. Hodson, H. K. and K. C. Hamner. 1970. Floral inducing extract
    from Xanthium. Science 167: 384-385.

17. Lam, S. L. 1965. Movement of the flower stimulus in Xanthium.
    Am. J. Botany 52: 924-928.

18. Leopold, A. C. and P. E. Kriedemann. 1975. Plant Growth and Development. 2nd Ed. McGraw-Hill Book Company, New York.

19. Lincoln, R. G., A. Cunningham, and K. C. Hammer. 1964. Evidence for a florigenic acid. Nature 202: 559-561.

20. Lincoln, R. G., D. L. Mayfield, and A. Cunningham. 1961. Preparation of a floral initiating extract from Xanthium. Science 133: 756.

21. Moore, T. C. 1979. Biochemistry and Physiology of Plant Hormones. Springer-Verlag, New York.

22. Ray, P. M. and W. E. Alexander. 1966. Photoperiodic adaptation to latitude in Xanthium strumarium. Am. J. Botany 53: 806-816.

23. Salisbury, F. B. 1958. The flowering process. Scient. Am. 198 (No. 4, Apr.): 109-117.

24. Salisbury, F. B. 1961. Photoperiodism and the flowering process. Ann. Rev. Plant Physiol. 12: 293-326.

25. Salisbury, F. B. 1963. The Flowering Process. Pergamon Press, Oxford.

26. Salisbury, F. B. 1965. The initiation of flowering. Endeavour 24: 74-80.

27. Salisbury, F. B. and C. W. Ross. 1978. Plant Physiology. 2nd Ed. Wadsworth Publishing Company, Belmont, California.

28. Searle, N. E. 1965. Physiology of flowering. Ann. Rev. Plant Physiol. 16: 97-118.

29. Takimoto, A. and K. C. Hammer. 1964. Effect of temperature and preconditioning on photoperiodic response of Pharbitis nil. Plant Physiol. 39: 1024-1030.

30. Wareing, P. F. and I. D. J. Phillips. 1970. The Control of Growth and Differentiation in Plants. Pergamon Press, New York.

31. Zeevaart, J. A. D. 1962. Physiology of flowering. Science 137: 723-731.

Special Materials and Equipment Required
Per Laboratory Section

(Approximately 100 to 150) Mature, dry fruits (burs) of cocklebur (Xanthium strumarium ≡ X. pennsylvanicum). Mature fruits of this common weed species can be collected locally in many areas

seasonally; alternatively, sufficient seeds to grow a small crop for
seed-increase purposes often can be obtained from an investigator
engaged in research on the physiology of flowering in cocklebur.
(72) 4-inch pots
(1) Greenhouse equipped with artificial lights and timer
(1) Light-tight cabinet or darkroom
(1) Camera with flash attachment and daylight-type flash bulb

Recommendations for Scheduling

This experiment is most satisfactorily conducted as a class project
in each laboratory section. Because of the relatively long duration of
the experiment, it is necessary to plant the seeds very early in the
course or even before the course begins.

Duration: Approximately 10 weeks; approximately 30 minutes required
during first laboratory period, about 15 to 30 minutes required once per
week each succeeding week, and approximately 1 to 1.5 hours required
during the laboratory period when the experiment is terminated.

EXERCISE 21
PHOTOPERIODIC CONTROL OF FLOWERING IN COCKLEBUR
(XANTHIUM STRUMARIUM)

REPORT

Name _____ Section _____ Date _____

Results of photoperiodic treatments:

No. photo-inductive cycles	No. plants in group	No. plants flowering	% flowering	Total fresh weight of floral parts of all plants in group (g)	Average fresh weight of floral parts per flowering plant (g)
0	_____	_____	_____	_____	_____
1	_____	_____	_____	_____	_____
2	_____	_____	_____	_____	_____
4	_____	_____	_____	_____	_____
8	_____	_____	_____	_____	_____
1-interrupted	_____	_____	_____	_____	_____

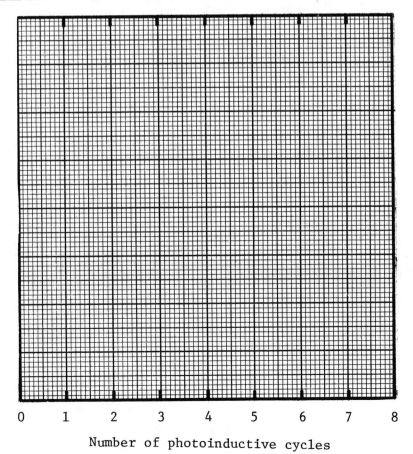

Number of photoinductive cycles

PHOTOPERIODIC CONTROL OF FLOWERING IN COCKLEBUR
(XANTHIUM STRUMARIUM)

REPORT

Name _____ Section _____ Date _____

Questions

1. Explain the effect of interrupting the single inductive dark period for group 6.

2. Define "ripeness to flower," and characterize this condition in the cocklebur.

3. Do your data provide evidence for a cumulative effect of photoinduction in the cocklebur? Explain.

4. Would you expect your results to be different for the plants receiving two or more photoinductive cycles if the cycles had been alternated singly with non-inductive cycles consisting of 20 hours of light and 4 hours of darkness? Explain.

EXERCISE 21

PHOTOPERIODIC CONTROL OF FLOWERING IN COCKLEBUR
(XANTHIUM STRUMARIUM)

REPORT

Name _____ Section _____ Date _____

5. What is the most critical environmental requirement for floral
    initiation in short-day plants? In long-day plants?

6. Is high-intensity light required for flowering in the cocklebur?
    Discuss.

7. Discuss the evidence for the proposition that floral initiation in
    cocklebur is mediated by a florigen.

MINERAL NUTRITION OF SUNFLOWERS

Introduction

Autotrophic plants generally require--in addition to carbon, hydrogen
and oxygen which they obtain from carbon dioxide, water and molecular
oxygen--thirteen elements which they absorb as inorganic ions (Table 22-I).
Six of these elements are required in greater amounts than the others and
are called "macronutrient" or "major" elements. They are nitrogen, phosphorus, potassium, sulfur, calcium and magnesium. The seven "micronutrient,"
"minor," or "trace" elements are iron, manganese, boron, copper, zinc,
chlorine and molybdenum. Several elements are required by some species but
not by others. For example, sodium is required for certain blue-green
algae and halophytes (e.g. Atriplex vesicara). Cobalt is a micronutrient
for some microorganisms and symbionts, although it has not been demonstrated to be essential for green plants. Silicon is indispensable for

Table 22-I. Essential Elements Absorbed as Inorganic
Ions by Most Autotrophic Plants[1]

Element	Symbol	Ion(s) most commonly absorbed	Approx. conc. in dry tissues (ppm)	Approx. optimum conc. in nutrient solution (mg atoms/liter)
Micronutrient elements				
Molybdenum	Mo	$MoO_4^=$	0.1	0.0001
Copper	Cu	Cu^+, Cu^{++}	6	0.0003
Zinc	Zn	Zn^{++}	20	0.001
Manganese	Mn	Mn^{++}	50	0.01
Iron	Fe	Fe^{++}, Fe^{+++}	100	0.1
Boron	B	$BO_3^=$, $B_4O_7^=$	20	0.05
Chlorine	Cl	Cl^-	100	0.05
Macronutrient elements				
Nitrogen	N	NO_3^-, NH_4^+	15,000	15
Potassium	K	K^+	10,000	5
Calcium	Ca	Ca^{++}	5,000	3
Phosphorus	P	$H_2PO_4^-$, $HPO_4^=$	2,000	2
Magnesium	Mg	Mg^{++}	2,000	1
Sulfur	S	$SO_4^=$	1,000	1

[1]Data mainly from P. R. Stout, Proc. 9th Ann. Calif. Fertilizer Conf.,
pp. 21-23, 1961 in Plant Physiology, 2nd Ed., by Frank B. Salisbury and Cleon W.
Ross, 1978. Wadsworth Publishing Company, Inc., Belmont, California 94002.

Table 22-II. Composition of Nutrient Solutions[2]

Stock Solution	1 Complete	2 -N	3 -K	4 -P	5 -Ca	6 -Mg	7 -S	8 -Fe	9 -6 Micronutrients
1 M Ca(NO$_3$)$_2$	10 ml	- ml	10 ml	10 ml	- ml	10 ml	10 ml	10 ml	10 ml
1 M KNO$_3$	10	-	-	10	10	10	10	10	10
1 M MgSO$_4$	4	4	4	4	4	-	-	4	4
1 M KH$_2$PO$_4$	2	2	-	-	2	2	2	2	2
FeEDTA[3]	2	2	2	2	2	2	2	-	2
Micronutrients[4]	2	2	2	2	2	2	2	2	-
1 M NaNO$_3$	-	-	10	-	20	-	-	-	-
1 M MgCl$_2$	-	-	-	-	-	-	4	-	-
1 M Na$_2$SO$_4$	-	-	-	-	-	4	-	-	-
1 M NaH$_2$PO$_4$	-	-	2	-	-	-	-	-	-
1 M CaCl$_2$	-	10	-	-	-	-	-	-	-
1 M KCl	-	10	-	2	-	-	-	-	-

[2]Modified from Plants in Action: A Laboratory Manual of Plant Physiology by Leonard Machlis and John G. Torrey. W. H. Freeman and Company. Copyright © 1956.

[3]FeEDTA is an iron-chelate complex, specifically sodium ferric ethylenediaminetetraacetate ("Sequestrene NaFe Iron Chelate" of Geigy Chemical Corporation, which contains 12% metallic iron.) Each milliliter of stock solution contains 5 mg of metallic iron ($\equiv$ 42 mg of the commercial chelate).

[4]Micronutrient stock solution (minus iron) contains per liter: 2.86 g H$_3$BO$_3$, 1.81 g MnCl$_2$ · 4H$_2$O, 0.11 g ZnCl$_2$, 0.05 g CuCl$_2$ · 2H$_2$O, and 0.025 g Na$_2$MoO$_4$ · 2H$_2$O.

diatoms, and vanadium is reported to be essential for one green alga (<u>Scenedesmus</u> <u>obliquus</u>).

Twelve of the thirteen elements in Table 22-I are derived from parent rock and are therefore "mineral elements." The ultimate source of nitrogen is, of course, molecular nitrogen (N_2) of the earth's atmosphere. However, aside from those plants which fix atmospheric nitrogen, either by themselves of symbiotically, nitrogen is absorbed as an inorganic ion (nitrate or ammonium ion) by autotrophic plants. Thus nitrogen usually is included in discussions of the "mineral nutrition" of plants.

One approach to determining the metabolic role of an essential element is to determine the consequence of its deficiency. In the case of all the macronutrient elements, there are characteristic symptoms which develop in the shoots of deficient plants. This type of approach often yields sufficient information to suggest metabolic blocks and thus specific sites where a particular element may function. The common method of implementing this approach is to grow plants hydroponically in solutions of precisely known chemical composition.

The purposes of this experiment are to become familiar with a method of growing plants hydroponically, learn deficiency symptoms associated with inadequate supplies of particular essential elements, and investigate some consequences of ion uptake by plants from unbalanced nutrient solutions.

Materials and Methods

Each group of students will be provided with 10 half-gallon Mason jars wrapped tightly with aluminum foil and bearing tags denoting the solutions they are to contain. Supplied with each jar will be either a flat cork or a metal lid perforated with three holes of approximately 1 cm diameter. Wash each jar thoroughly with detergent and hot water, rinse with tap water, and finally rinse with distilled water. Be careful not to tear the aluminum foil and to patch any torn places that may be present. Similarly wash the cork or metal lid; if the latter is used, dip the disk part of the lid in molten paraffin before it is used.

Fill each of the jars about two-thirds full of distilled water, and prepare the solution in each jar which is denoted by the tag. Give the tenth jar, labeled "Unknown," to the instructor, who will prepare that solution. Add, one at a time, the volumes of particular stock solutions, as specified in Table 22-II. Stir the solution with a clean glass rod upon addition of each of the components, so as to prevent precipitation reactions. When all 9 of the solutions indicated in Table 22-II have been prepared, fill each jar to the neck with distilled water and stir thoroughly.

Measure and record the pH of the "Complete" and "-N" solutions.

Carefully remove 9- to 12-day old sunflower seedlings from a flat

257

of vermiculite or sand, avoiding as much as possible breaking of roots. Remove all clinging debris and rinse the root systems of the seedlings once in tap water and then once in distilled water. Using a forceps to assist in guiding the roots through the holes of the jar lid or cork, mount three seedlings in each jar. Support each seedling in the lid or cork with a gasket of cotton (Be careful not to injure the plant!) around the hypocotyl. When properly mounted, the seedling shoots should extend about 5 cm above the top of the jar, and the remaining hypocotyl-root system should be freely suspended in the solution. The cotton gaskets must not be in contact with the solution.

Place all the plants in a greenhouse or controlled-environment facility where they will be subjected to a 12- to 16-hour photoperiod at a temperature of approximately 20 to 29° C and a light intensity of 1,500 to 3,000 ft-c and an 8- to 12-hour dark period at approximately 18 to 23° C.

Observe the plants at each laboratory period. Check the level of solution in each jar and replenish the water lost by adding distilled water. Two or three weeks after the start of the experiment, describe the deficiency symptoms which have developed, again measure and record the pH of the "Complete" and "-N" solutions, and measure the lengths of the shoots of the plants in each group. Record the data from your group, and/or the compiled data for the class as directed by the instructor, on the data report forms provided.

References

1. Bonner, J. and A. W. Galston. 1952. Principles of Plant Physiology. W. H. Freeman and Company, San Francisco.

2. Bollard, E. G. and G. W. Butler. 1966. Mineral nutrition of plants. Ann. Rev. Plant Physiol. 17: 77-112.

3. Brouwer, R. 1965. Ion absorption and transport in plants. Ann. Rev. Plant Physiol. 16: 241-266.

4. Broyer, T. C. and P. R. Stout. 1959. The macronutrient elements. Ann. Rev. Plant Physiol. 10: 277-300.

5. Epstein, E. 1972. Mineral Nutrition of Plants: Principles and Perspectives. John Wiley and Sons, New York.

6. Evans, H. J. and G. J. Sorger. 1966. Role of mineral elements with emphasis on the univalent cations. Ann. Rev. Plant Physiol. 17: 47-76.

7. Gauch, H. G. 1972. Inorganic Plant Nutrition. Dowden, Hutchinson, and Ross, Stroudsburg, Pennsylvania.

8. Hoagland, D. R. and D. I. Arnon. 1950. The Water-culture Method for Growing Plants Without Soil. Calif. Agric. Exp. Sta. Circular 347.

9. Meyer, B. S., D. B. Anderson, and R. H. Böhning. 1960. Introduction to Plant Physiology. D. Van Nostrand Company, New York.

10. Ray, P. M. 1972. The Living Plant. 2nd Ed. Holt, Rinehart and Winston, New York.

11. Salisbury, F. B. and C. W. Ross. 1978. Plant Physiology. 2nd Ed. Wadsworth Publishing Company, Belmont, California.

12. Schütte, K. H. 1964. The Biology of the Trace Elements. Their Role in Nutrition. J. B. Lippincott Company, Philadelphia.

13. Steward, F. C. 1963. Plant Physiology. Vol. III. Inorganic Nutrition of Plants. Academic Press, New York.

14. Stewart, I. 1963. Chelation in the absorption and trans-location of mineral elements. Ann. Rev. Plant Physiol. 14: 295-310.

15. Stiles, W. 1961. Trace Elements in Plants. University Press, Cambridge.

16. Sutcliffe, J. F. 1962. Mineral Salts Absorption in Plants. Pergamon Press, New York.

17. Wallace, T. 1961. The Diagnosis of Mineral Deficiencies in Plants. 3rd Ed. Chem. Publ. Company, New York.

18. Webster, G. L. 1959. Nitrogen Metabolism in Plants. Row, Peterson and Company, Evanston, Illinois.

Special Materials and Equipment Required
Per Team of 3-8 Students

(30 to 50) Sunflower (Helianthus annuus) seedlings 9 to 12 days old
(10) Half-gallon size Mason jars wrapped light-tight with aluminum foil
(10) Metal lids or flat corks, each perforated with 3 holes about 1 cm in diameter, to fit jars.
(Approximately 1/4 lb) Household paraffin
(1 package) Cotton
(Stock supply) Distilled water
(1) Greenhouse or controlled-environment growth chamber
(1) pH meter
(100 ml) 1 M $Ca(NO_3)_2$
(100 ml) 1 M KNO_3
(50 ml) 1 M $MgSO_4$
(25 ml) 1 M KH_2PO_4
(25 ml) FeEDTA stock solution, prepared as described in Materials and Methods

(25 ml) Micronutrients stock solution, prepared as described in Materials
    and Methods
(50 ml) 1 M NaNO$_3$
(10 ml) 1 M MgCl$_2$
(10 ml) 1 M Na$_2$SO$_4$
(10 ml) 1 M NaH$_2$PO$_4$
(25 ml) 1 M CaCl$_2$
(25 ml) 1 M KCl

Recommendations for Scheduling

It is recommended that at least two teams in each laboratory section
set up a complete experiment. Alternatively, individual students or small
groups of students may be assigned to prepare specified solutions in
duplicate and the entire experiment be treated as a class project. Sunflower seedlings should be grown in advance and available at the laboratory when the experiment is to be started.

Duration: 2 to 3 weeks; approximately 1.5 hours required during
period experiment is started, 15 to 30 minutes required at least once
weekly thereafter, and 1.5 to 2 hours required during period when
experiment is terminated.

EXERCISE 22

MINERAL NUTRITION OF SUNFLOWERS

REPORT

Name _____ Section _____ Date _____

Changes in pH in nutrient solutions:

Solution	Initial pH	Final pH
Complete	_____	_____
-N	_____	_____

Comparative shoot growth:

Solution	Mean final shoot length (cm)
Complete	_____
-N	_____
-K	_____
-P	_____
-Ca	_____
-Mg	_____
-S	_____
-Fe	_____
-6 micro-nutrients	_____
Unknown	_____

MINERAL NUTRITION OF SUNFLOWERS

REPORT

Name _____   Section _____   Date _____

Descriptions of deficiency symptoms:

-Nitrogen:

-Potassium:

-Phosphorous:

-Calcium:

EXERCISE 22

MINERAL NUTRITION OF SUNFLOWERS

REPORT

Name _____ Section _____ Date _____

<u>Descriptions of deficiency symptoms - continued</u>:

   -<u>Magnesium</u>:

   -<u>Sulfur</u>:

   -<u>Iron</u>:

   -<u>6 micronutrients</u>:

MINERAL NUTRITION OF SUNFLOWERS

REPORT

Name _____   Section _____   Date _____

<u>Identification of unknown nutrient solution and reasons:</u>

EXERCISE 22

MINERAL NUTRITION OF SUNFLOWERS

REPORT

Name _____ Section _____ Date _____

Questions

1. Explain the observed changes, if any, in pH in the "Complete" and "-N" solutions.

2. What is a chelate, and why was iron supplied as a chelated complex rather than as a simple salt such as $FeCl_3$?

3. Can any of the symptoms displayed by the plants growing in the "-6 micronutrients" solutions reasonably be attributed to deficiencies of specific essential elements? Explain.

MINERAL NUTRITION OF SUNFLOWERS

REPORT

Name _____ Section _____ Date _____

4. In general, how similar would the deficiency symptoms exhibited by the sunflower plants be to symptoms in other seed plants?

5. Of what practical value or economic importance is hydroponics currently?

EXERCISE 23

POTASSIUM ACTIVATION OF PYRUVIC KINASE

Introduction

Lucanus in 1865 and Birner and Lucanus in 1866 first provided conclusive evidence that potassium (K) is essential for oat plants (Avena sativa). Since that important discovery, K has been proven to be required by a wide variety of higher plants and other organisms, and it can be confidently concluded that the element is generally indispensable for all living organisms.

A review of available information published by H. J. Evans and G. J. Sorger in 1966 revealed that K^+ and other univalent cations (Rb^+, NH_4^+, Na^+, Li^+) function as cofactors (activators) for some 46 enzymes from animals, plants and microorganisms. It appears that both the magnitude of the K^+ requirement and its specificity as an essential mineral element can be fully accounted for, at least in the case of higher plants, on the basis of the activity of the cation as a cofactor for enzymes.

Even though K^+ is specifically required by all plants, as well as other organisms, it has been observed that some other univalent cations (e.g. Rb^+ and Na^+) can partially substitute for K^+ in certain plants when the supply of K^+ is limiting. The stimulation of growth by Rb^+ and Na^+ of many plants deprived of an optimum supply of K^+ is correlated with the capacity of these cations to activate, to some extent at least, some but not all enzymes requiring univalent cations. When the effects of different univalent cations on activity of specific enzymes are examined, they are found to differ markedly. Thus both K^+ and Rb^+ may activate a particular enzyme, and Na^+ and Li^+ may fail to activate or even inhibit the enzyme. Potassium appears to be the only element available in nature in sufficient quantity and with the appropriate chemical properties to adequately activate the majority of univalent cation-activated enzymes.

Among the numerous enzymes known to be activated by univalent cations is pyruvic kinase, which catalyzes the following reversible reaction:

$$\text{Phosphoenolpyruvate + ADP} \underset{Mg^{++} \text{ or } Mn^{++}, K^+}{\overset{\text{Pyruvic kinase}}{\rightleftharpoons}} \text{Pyruvate + ATP}$$

Pyruvic kinase from pea (Pisum sativum) is activated most effectively by K^+ but is stimulated also, in decreasing order of effectiveness, by Rb^+ > NH_4^+ > Na^+. The enzyme also has an absolute divalent cation requirement that is satisfied by Mg^{++} or Mn^{++}.

The purpose of this experiment (actually two experiments) is to investigate the effects of K^+ and some other univalent cations on the activity of a highly purified preparation of pyruvic kinase in vitro.

267

Materials and Methods[1]

Each team of students will perform one of the two experiments described. However, each student should report and interpret the data for both experiments. In both experiments, pyruvic kinase will be assayed by the determination of ADP-dependent formation of pyruvate. The yield of pyruvate will be measured by first reacting the α-keto-acid with 2,4-dinitrophenylhydrazine reagent and then measuring colorimetrically the amount of the pyruvic acid 2,4-dinitrophenylhydrazone which forms.

The first operation for each team of students is to prepare a standard curve for the colorimetric assay. Prepare a series of 12 test tubes, with two tubes each containing 1.0 ml of pyruvic acid solution at one of the following concentrations: 0, 0.1, 0.2, 0.4, 0.6 and 0.8 mM. Add 1.0 ml of 0.0125% 2,4-dinitrophenylhydrazine in 2 N HCl (Caution: Do not pipet by mouth!), and mix each tube thoroughly. Incubate the tubes for 10 minutes at 37° C. Then remove the tubes from the water bath, add to each tube 2.0 ml of water and 5.0 ml of 0.6 N NaOH and allow them to stand for 10 minutes at room temperature for color development. Then measure the absorbance at 510 nm in a colorimeter or spectrophotometer. Plot a standard curve.

Experiment 1. Effects of different salts of K$^+$. Prepare 5 kinds of reaction mixtures, each kind in duplicate, in 15-ml conical, glass-stoppered tubes (or other small tubes which can be centrifuged in a clinical centrifuge). Add to each tube the components indicated in Table 23-I, in the order in which they are listed. Incubate the reaction mixtures in a water bath at 37° C for 10 minutes. Stop the reactions by adding 1.0 ml of cold 0.0125% 2,4-dinitrophenylhydrazine reagent (Caution!). Incubate for an additional 10 minutes in the 37° C bath. Then remove the tubes from the bath and add to each 2.0 ml of water and 5.0 ml of 0.6 N NaOH. Allow the tubes to stand for 10 minutes at room temperature. Then centrifuge the reaction mixtures at top speed in a clinical centrifuge for 5 to 10 minutes to precipitate protein or Mg(OH)$_2$. Measure the absorbance (O.D.) at 510 nm in a colorimeter or spectrophotometer. Use the supernatant fractions from tubes 9 and 10 as blanks. Record the average absorbance values and the calculated (see standard curve) amounts of pyruvate formed in a table.

Experiment 2. Effects of chlorides of different univalent cations. Prepare two reaction mixtures exactly as described for tubes 1 and 2 in Experiment 1. Also prepare two reaction mixtures each containing:

(a) 0.1 ml of 0.5 M NaCl instead of KCl
(b) 0.1 ml of 0.5 M LiCl instead of KCl
(c) 0.1 ml additional Tris buffer instead of KCl

---

[1]Adapted from: Miller G. and H. J. Evans. 1957. The influence of salts on pyruvate kinase from tissues of higher plants. Plant Physiol. 32: 346-354.

Table 23-I. Composition of Reaction Mixtures for Experiment 1

Stock solution	Volume per reaction mixture (ml)				
	Tubes 1-2[4]	Tubes 3-4	Tubes 5-6	Tubes 7-8	Tubes 9-10
Tris-HCl buffer, 0.1 M, pH 7.4	0.5	0.5	0.5	0.6	0.6
Cyclohexylammonium salt of PEP, 0.015 M	0.1	0.1	0.1	0.1	0.1
Tris ADP, 0.025 M[2]	0.1	0.1	0.1	0.1	--
$MgSO_4$, 0.1 M	0.1	0.1	0.1	0.1	0.1
KCl, 0.5 M	0.1	--	--	--	0.1
KNO_3, 0.5 M	--	0.1	--	--	--
K_2SO_4, 0.25 M	--	--	0.1	--	--
Dialyzed, diluted enzyme solution[3]	0.1	0.1	0.1	0.1	0.1

[2]Tris ADP can be purchased as such or it can be prepared from the barium salt of ADP by a procedure described in reference 13.

[3]A highly purified commercial preparation (solution) of pyruvic kinase is dialyzed approximately 6 to 12 hours against Tris-HCl buffer (0.05 M, pH 7.4) and diluted before use with Tris buffer so that 0.1 ml contains sufficient activity in reaction mixtures 1 and 2 to catalyze the formation of approximately 0.5 μmole of pyruvate in 10 minutes at 37° C.

[4]The actual amounts of the components in reaction mixtures 1 and 2 are: 50 μmoles of Tris buffer, 1.5 μmoles of cyclohexylammonium salt of PEP, 2.5 μmoles of Tris ADP, 10 μmoles of $MgSO_4$, and 50 μmoles of KCl, all in a total final volume of 1.0 ml.

Run the reactions and perform the colorimetric assays as described for Experiment 1. Record the means of the observed absorbance values and the calculated amounts of pyruvate formed in a table.

References

1. Bernhard, S. 1968. The Structure and Function of Enzymes. W. A. Benjamin, New York.

2. Conn, E. E. and P. K. Stumpf. 1963. Outlines of Biochemistry. 2nd Ed. John Wiley and Sons, New York.

3. Cleland, W. W. 1967. Enzyme kinetics. Ann. Rev. Biochem. 36: 77-112.

4. Dixon, M. and E. C. Webb. 1964. Enzymes. 2nd Ed. Academic Press, New York.

5. Epstein, E. 1972. Mineral Nutrition of Plants: Principles and Perspectives. John Wiley and Sons, New York.

6. Evans, H. J. 1963. Effect of potassium and other univalent cations on activity of pyruvate kinase in Pisum sativum. Plant Physiol. 38: 397-402.

7. Evans, H. J. and G. J. Sorger. 1966. Role of mineral elements with emphasis on the univalent cations. Ann. Rev. Plant Physiol. 17: 47-76.

8. Gauch, H. G. 1972. Inorganic Plant Nutrition. Dowden, Hutchinson, and Ross, Stroudsburg, Pennsylvania.

9. Giese, A. C. 1968. Cell Physiology. 3rd Ed. W. B. Saunders Company, Philadelphia.

10. Kachmar, J. F. and P. D. Boyer. 1953. Kinetic analysis of enzyme reactions. II. The K^+ activation and Ca^{++} inhibition of pyruvic phosphoferase. J. Biol. Chem. 200: 669-682.

11. Lehninger, A. L. 1975. Biochemistry. 2nd Ed., Worth Publishers, New York.

12. Mahler, H. R. and E. H. Cordes. 1971. Biological Chemistry. 2nd Ed. Harper and Row Publishers, New York.

13. Miller, G. and H. J. Evans. 1957. The influence of salts on pyruvate kinase from tissues of higher plants. Plant Physiol. 32: 346-354.

14. Ray, P. M. 1972. The Living Plant. 2nd Ed. Holt, Rinehart and Winston, New York.

15. Salisbury, F. B. and C. W. Ross. 1978. Plant Physiology. 2nd Ed. Wadsworth Publishing Company, Belmont, California.

16. Steward, F. C. 1963. Plant Physiology. Vol. III. Inorganic Nutrition of Plants. Academic Press, New York.

Special Materials and Equipment Required
Per Team of 3-8 Students

(1 ml) Pyruvic kinase solution, previously dialyzed and diluted as described in Materials and Methods
(6 ml) Tris-HCl buffer, 0.1 M, pH 7.4
(1 ml) Solution of cyclohexylammonium salt of PEP, 0.015 M
(1 ml) Tris ADP solution, 0.025 M
(1 ml) 0.1 M $MgSO_4$
(1 ml) 0.5 M KCl
(1 ml) 0.5 M KNO_3 (Experiment 1)
(1 ml) 0.25 M K_2SO_4 (Experiment 1)
(1 ml) 0.5 M NaCl (Experiment 2)
(1 ml) 0.5 M LiCl (Experiment 2)
(2 ml) Pyruvic acid solution at each of the following concentrations: 0, 0.1, 0.2, 0.4, 0.6 and 0.8 mM
(25 ml) 2,4-Dinitrophenylhydrazine reagent (0.0125% in 2 N HCl)
(150 ml) 0.6 N NaOH
(12) Test tubes to fit rotor of clinical centrifuge
(1) Thermoregulated water bath
(1) Colorimeter or spectrophotometer

Recommendations for Scheduling

Each team of 3 to 8 students should perform one complete experiment, including preparation of a standard curve for the colorimetric assay. In order for the experiment to be completed in a single laboratory period, it is necessary that the enzyme solution and all reagents be prepared in advance. If laboratory periods are less than 3 hours long, it may be necessary also to furnish students with a standard curve and restrict their assignment to only that part of the experiment involving the enzyme assay.

Duration: 1 entire laboratory period.

POTASSIUM ACTIVATION OF PYRUVIC KINASE

REPORT

Name _____ Section _____ Date _____

Standard curve for colorimetric assay:

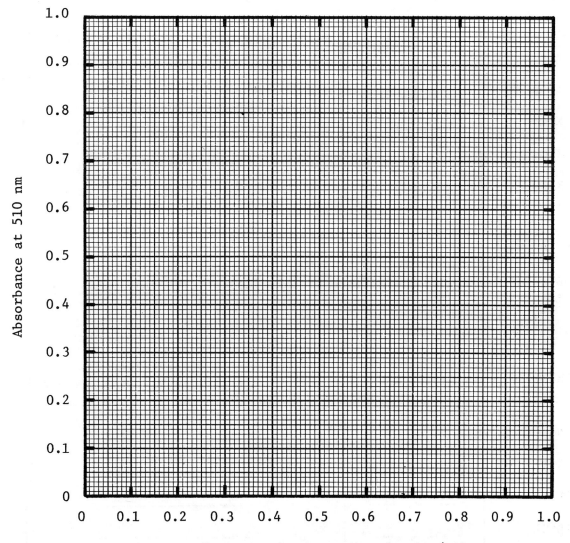

Pyruvate concentration (µmoles/ml)

POTASSIUM ACTIVATION OF PYRUVIC KINASE

REPORT

Name _____ Section _____ Date _____

Results of Experiment 1 – Effects of different salts of K^+:

Salt of K^+	Absorbance[1] (510 nm)	Pyruvic kinase activity[1] (μmoles pyruvate formed in 10 minutes)
None	_____	_____
KCl	_____	_____
KNO$_3$	_____	_____
K$_2$SO$_4$	_____	_____

[1]Average of _____ determinations.

Results of Experiment 2 – Effects of chlorides of different univalent cations:

Chloride	Absorbance[1] (510 nm)	Pyruvic kinase activity[1] (μmoles pyruvate formed in 10 minutes)
None	_____	_____
KCl	_____	_____
NaCl	_____	_____
LiCl	_____	_____

[1]Average of _____ determinations.

EXERCISE 23

POTASSIUM ACTIVATION OF PYRUVIC KINASE

REPORT

Name _____ Section _____ Date _____

Questions

1. What was the specific purpose of Experiment 1, and what conclusions
    are warranted on the basis of the results of this experiment?

2. Summarize the comparative effectiveness of K^+, Na^+ and Li^+, as judged
    from the results of Experiment 2.

3. What was the source (organism, tissue) of the pyruvic kinase used in
    these experiments? Would you expect the results of Experiment 2 to be
    different with pyruvic kinase from other sources? Explain.

4. Discuss briefly the mechanism by which K^+ and other univalent cations
    are believed to effect enzyme activation.

POTASSIUM ACTIVATION OF PYRUVIC KINASE

REPORT

Name _____ Section _____ Date _____

5. Of what major catabolic pathway is the reaction investigated in these
    experiments a part? What term is given to the specific type of ATP
    formation which occurs in this reaction?

6. Explain the necessity of divalent Mg^{++} or Mn^{++} ions for the pyruvic
    kinase reaction.

EXERCISE 24

ABSORPTION AND TRANSLOCATION OF PHOSPHATE-^{32}P

Introduction

Terrestrial seed plants require approximately 12 mineral elements, in addition to nitrogen, which they absorb as inorganic ions through their roots from the soil. Ion absorption occurs by two basic processes common-ly known as passive and active ion uptake. Passive absorption is largely a diffusional process and does not require direct expenditure of metabolic energy. Active absorption, in contrast, is a metabolically dependent, highly regulated process by which certain species of ions are accumulated in the plant to concentrations greatly in excess of the concentration in the soil, while other ion species are excluded. Both passive and active ion uptake processes generally are influenced by factors which affect respiration and other metabolic processes in the root, with the latter process being more directly affected in this way than the former. Factors which influence transpiration rate often influence rates of ion absorption also, but passive ion absorption is more directly subject to influence in this way than is active uptake.

Once absorbed ions have traversed the exterior tissues of the root to the stele, they are translocated predominantly via the conducting elements of the xylem upward to the shoot. Some leakage from xylem to phloem occurs, however, so that the xylem may not be the exclusive tissue of initial transit of some ions.

Upward translocation of ions in the xylem is influenced markedly by factors which affect the transpiration rate. Thus ions in the xylem sap generally move at rates corresponding to the rate of the bulk flow of water, which in turn is directly correlated with the rate of transpiration when the latter process is occurring. During periods of active transpira-tion, passive absorption of ions also is accelerated mainly because the relatively rapid bulk movement of xylem sap maintains a steep diffusional gradient from soil to stele. When transpiration is not occurring, move-ment of ions upward in the xylem sap still continues by the diffusion of individual ions within the xylem sap.

Following initial translocation upward via the xylem, some ions (e.g. Ca^{++} and Fe^{++} or Fe^{+++}) become essentially permanently bound in that part of the plant to which they are initially transported. Such elements are said to be immobile in the plant; that is, they tend not to be redistribu-ted. Other essential elements (e.g. N, P and K) are relatively mobile. For example, a phosphate ion may initially enter a leaf via the xylem, become assimilated into an organic molecule in the leaf, subsequently be liberated as a small molecule or ion in catabolic processes, then be exported from the leaf via the phloem and be translocated, either upward or downward, depending on the position of the leaf, to some other part of the plant. Since some movement between xylem and phloem occurs, the mobile elements thus tend actually to be circulated within the plant.

The purpose of this experiment is to investigate the effects of certain physical environmental factors on the absorption and translocation

of phosphorus in a selected dicot. Actually, the net result of both
absorption and translocation will be measured by monitoring the accumulation of radioactive ^{32}P in a leaf of a plant which is supplied with
phosphate-^{32}P via the roots in a mineral nutrient solution.

Materials and Methods

Very carefully remove an approximately 3- to 4-week-old bean, sunflower or other dicot from the substrate in which it is rooted, avoiding
as much as possible breaking or injuring the roots. Gently wash the root
system with tap water to remove debris and rinse with distilled water.

Place the root system in a small beaker (e.g. 400-ml) and promptly
add the minimum volume of complete mineral nutrient solution[1] necessary
to completely submerge the roots (e.g. 250-ml). Then add to the nutrient
solution 10 to 25 microcuries (µc) of orthophosphate-^{32}P ($H_3{}^{32}PO_4$).
Caution: Do not pipet by mouth; follow the special directions for handling of radioisotopes given by the instructor. Stir the solution thoroughly. If possible, bubble air or oxygen through the nutrient solution
throughout the experimental period.

Anchor the shoot of the plant to an adjacent ring stand. By means
of small strips of tape on the margin, flatten and fasten an upper half-
to two-thirds grown leaf or leaflet to a vertically mounted piece of lead
sheet, plexiglass or other material. The purpose of this is to hold the
leaf in a stationary position and to shield it from the rest of the plant.

While the plant is adjusting to the nutrient solution, determine the
background radiation in the laboratory, using the Geiger-Müller detector
and scaler provided and following the directions given by the instructor.
Record the background count rate (counts per minute, cpm).

Now mount the Geiger-Müller detector tube in a fixed position per-
pendicular to the mounted leaf blade, with the window of the detector
positioned only 1 to 2 mm from the leaf surface. Throughout the remainder
of the laboratory period, preferably for a period of several hours, record
the accumulated counts and count rate in the mounted leaf at 15-minute
intervals. Record all data in a table on a chalkboard in the laboratory.

When prepared to start monitoring accumulation of ^{32}P in the mounted
leaf, expose the plant to high-intensity light provided by a 150-watt
floodlight held in a photographic reflector mounted on an adjacent ring
stand. Do not position the lamp so close as to cause excessive heating
of the plant. Monitor accumulation of ^{32}P in the mounted leaf for 2 to
3 hours.

If arrangements can be made to continue the experiment beyond the

---

[1]See Exercise 22 - "Mineral Nutrition of Sunflowers" for composition
of nutrient solution.

scheduled laboratory period, change the environmental conditions to darkness. Either cover the entire plant and accessory apparatus with a large box or set it in a light-tight cabinet, leaving the scaler exposed. Continue to monitor accumulation of ^{32}P in the mounted leaf for the succeeding 2 to 3 hours.

Following the period in darkness, if possible, change environmental conditions again. First, again expose the shoot to high-intensity light for one hour. Then cool the nutrient solution to approximately 10° C (by placing the beaker in an ice-water bath), and continue to monitor accumulation of ^{32}P for an additional 2 to 3 hours.

Record all data in the table provided in the Report forms. Plot two curves on a common set of coordinates, one depicting the net count rate versus time, and the other, the total net counts accumulated versus time.

A very interesting alternative or optional addition to this experiment is autoradiography of the whole plant at the end of the experiment described previously, or simply after permitting a plant to absorb and translocate the isotope for approximately two to several hours in light.

The technique of autoradiography is analogous to photography, except that the ionizing radiation (high-energy beta particles in the case of ^{32}P) rather than photons of light "exposes" the film. Commonly autoradiography is done with x-ray film, which, when subsequently developed, yields a dark image on a transparent background. The product of this process is an autoradiogram (or radioautogram).

To perform autoradiography, first remove the plant from the nutrient solution and rinse the root system thoroughly with tap water. Place the plant between newspaper and blotters in a plant press (such as routinely used by taxonomists) and dry the plant at approximately 75° C in a forced draft oven. If the plant is rather large, bend the shoot and roots, as it is being pressed, so that the dried specimen will fit on a 10 x 12 inch sheet of film.

When the plant is dry, take it to a darkroom, along with suitable 10 x 12 inch film (e.g. Kodak Duplitized No-screen Medical X-ray film) and a film holder (e.g. Kodak X-ray Exposure holder, for 10 x 12 inch film). Mount the dried plant directly on the emulsion of the film, enclose the film in the holder, place weight on top of the film holder to keep the specimen flat against the film, and, after an estimated suitable exposure time, develop the film as directed by the manufacturer.

It is possible to calculate an approximate exposure time, provided certain information is available (see reference 23). However, under the conditions of this experiment, exposure time can best be determined the hard way, that is by trial and error. It is recommended that the initial trial exposure time be 12 hours. If this fails to provide a good image, repeat the trial exposures for 3, 6, 12 and 18 days until a good image is produced. It is useful to note that, as in photography, the density of the image on the autoradiogram is proportional to the logarithm of the exposure rather than to the exposure per se. When a good autoradiogram is produced, prepare a photocopy of it and include the copy in your report.

References

1. Asher, C. J. and J. F. Loneragan. 1967. Response of plants to phosphate concentration in solution cultures. I. Growth and phosphorus content. Soil Sci. 103: 225-233.

2. Biddulph, O. 1955. Studies of mineral nutrition by use of tracers. Botan. Rev. 21: 251-295.

3. Biddulph, O., S. Biddulph, R. Cory, and H. Koontz. 1958. Circulation patterns for phosphorus, sulfur and calcium in the bean plant. Plant Physiol. 33: 293-300.

4. Biddulph, S. and O. Biddulph. 1959. The circulatory system of plants. Scient. Am. 200 (No. 2, Feb.): 44-49.

5. Bollard, E. G. 1960. Transport of the xylem. Ann. Rev. Plant Physiol. 11: 114-166.

6. Bollard, E. G. and G. W. Butler. 1966. Mineral nutrition of plants. Ann. Rev. Plant Physiol. 17: 77-112.

7. Briggs, G. E. 1967. Movement of Water in Plants. Davis Publishing Company, Philadelphia.

8. Brouwer, R. 1965. Ion absorption and transport in plants. Ann. Rev. Plant Physiol. 16: 241-266.

9. Crafts, A. S. 1961. Translocation in Plants. Hort, Rinehart and Winston, New York.

10. Epstein, E. 1972. Mineral Nutrition of Plants: Principles and Perspectives. John Wiley and Sons, New York.

11. Fried, M. and R. E. Shapiro. 1961. Soil-plant relationships in ion uptake. Ann. Rev. Plant Physiol. 12: 91-112.

12. Gauch, H. G. 1972. Inorganic Plant Nutrition. Dowden, Hutchinson, and Ross, Stroudsburg, Pennsylvania.

13. Jensen, R. D. and S. A. Taylor. 1961. Effect of temperature on water transport through plants. Plant Physiol. 36: 639-642.

14. Kozlowski, T. T. 1964. Water Metabolism in Plants. Harper & Row, Publishers, New York.

15. Ray, P. M. 1972. The Living Plant. 2nd Ed. Hort, Rinehart and Winston, New York.

16. Richardson, M. 1968. Translocation in Plants. St. Martin's Press, New York.

17. Rinne, R. W. and R. G. Langston. 1960. Studies on lateral movement of phosphorus 32 in peppermint. Plant Physiol. 35: 216-219.

18. Russell, R. S. and D. A. Barber. 1960. The relationship between salt uptake and the absorption of water by intact plants. Ann. Rev. Plant Physiol. 11: 127-140.

19. Salisbury, F. B. and C. W. Ross. 1978. Plant Physiology. 2nd Ed. Wadsworth Publishing Company, Belmont, California.

20. Sutcliffe, J. F. 1962. Mineral Salts Absorption in Plants. Pergamon Press, New York.

21. Sutcliffe, J. 1968. Plants and Water. St. Martin's Press, New York.

22. Wallace, A. 1966. Current Topics in Plant Nutrition. Edwards Bros., Ann Arbor, Michigan.

23. Wang, C. H. and D. L. Willis. 1965. Radiotracer Methodology in Biological Science. Prentice-Hall, Englewood Cliffs, New Jersey.

Special Materials and Equipment Required
Per Laboratory Section

(1) Vigorous 3- to 4-week-old bean, sunflower or other dicot plant
(10 to 25 microcuries) $H_3{}^{32}PO_4$
(Approximately 250 ml) Complete mineral nutrient solution (see Exercise 22 "Mineral Nutrition of Sunflowers" for composition)
(1) Scaler and Geiger-Müller detector
(1) Photographic reflector and 150-watt floodlight
(1) Large light-tight box or cabinet
(Approximately 5 sheets) Kodak Duplitized No-screen X-ray film, 10 x 12 inch
(1) Kodak X-ray Exposure Holder, for 10 x 12 inch film
(-) Reagents to develop X-ray film
(1) Plant press

Recommendations for Scheduling

It is recommended that this experiment be conducted as a demonstration. If possible, accumulation of ^{32}P in the mounted leaf should be monitored continuously at 15-minute intervals for 6 to 8 hours, with environmental conditions being changed as prescribed in the Materials and Methods section. The experiment can be started in a scheduled laboratory period and either be continued in another laboratory section on the same day or attended by students who volunteer to return to the laboratory at specified times after the regular scheduled period.

The optional autoradiography, if done, must be done either by the instructional staff or by students at times outside the regular scheduled laboratory period.

Duration: 6 to 8 hours (exclusive of autoradiography).

EXERCISE 24

ABSORPTION AND TRANSLOCATION OF PHOSPHATE-^{32}P

REPORT

Name _____ Section _____ Date _____

Accumulation of ^{32}P in mounted leaf:

Time counting started	Total counts	Duration of counting (minutes)	Gross counts per minute[1]	Background counts per minute[2]	Net counts per minute[3]	Total lapsed time (minutes)	Total net counts accumulated[4]

(Table continued on next page.)

EXERCISE 24

ABSORPTION AND TRANSLOCATION OF PHOSPHATE-^{32}P

REPORT

Name _____ Section _____ Date _____

Accumulation of ^{32}P in mounted leaf:

Time counting started	Total counts	Duration of counting (minutes[4])	Gross counts per minute[1]	Background counts per minute[2]	Net counts per minute[3]	Total lapsed time (minutes)	Total net counts accumulated[4]

(Table continued on next page.)

EXERCISE 24

ABSORPTION AND TRANSLOCATION OF PHOSPHATE-^{32}P

REPORT

Name _____ Section _____ Date _____

Accumulation of ^{32}P in mounted leaf:

Time counting started	Total Counts	Duration of counting (minutes)	Gross counts per minute[1]	Background counts per minute[2]	Net counts per minute[3]	Total lapsed time (minutes)	Total net counts accumulated[4]

[1] Total counts/duration of counting in minutes.

[2] Standard; determined at beginning of experiment.

[3] Gross counts per minute – background counts per minute.

[4] Sum of the net counts, computed separately for each measurement interval, for the total lapsed time recorded in the adjacent column.

EXERCISE 24

ABSORPTION AND TRANSLOCATION OF PHOSPHATE-^{32}P

REPORT

Name _____ Section _____ Date _____

Graph of counting data:

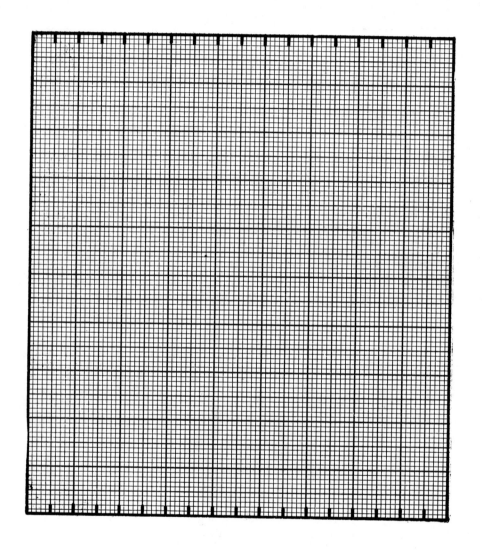

Net count rate (net counts accumulated/interval)

Total net counts accumulated

Time (minutes)

EXERCISE 24

ABSORPTION AND TRANSLOCATION OF PHOSPHATE-^{32}P

REPORT

Name _____ Section _____ Date _____

Copy of autoradiogram:

ABSORPTION AND TRANSLOCATION OF PHOSPHATE-^{32}P

REPORT

Name _____ Section _____ Date _____

Questions

1. Describe the kinetics of accumulation of ^{32}P in the mounted leaf during the initial environmental conditions of high-intensity light.

2. Describe the effect of each new set of conditions (transfer to darkness, cooling of root system and nutrient solution) on the rate of accumulation.

3. Explain the difference, if any, in the rate of accumulation of ^{32}P in light and darkness. Does the difference, if any, reflect mainly a difference in rate of <u>absorption</u> of phosphate in light and darkness, or a difference in rate of <u>translocation</u>? Explain carefully.

EXERCISE 24

ABSORPTION AND TRANSLOCATION OF PHOSPHATE-^{32}P

REPORT

Name _____ Section _____ Date _____

4. Was the effect, if any, of cooling primarily an effect on absorption or on translocation of phosphate? Explain.

5. Does the effect of temperature provide a clue as to whether phosphate ion absorption was passive or active? Explain.

6. In what ionic form(s) is phosphorus preferentially absorbed by most plants? If the pH of the nutrient solution was approximately 5, which ion of phosphorus was most abundant in the solution?

7. What did the autoradiogram reveal about the relative distribution of ^{32}P in the plant at the end of the experiment?

SYMBIOTIC NITROGEN FIXATION IN LEGUME NODULES

Introduction

Nitrogen is an essential element for all living organisms. Although N accounts for only 1 to 2% of the dry weight of most plant materials, it is exceeded in amount only by carbon, hydrogen and oxygen.

Nitrogen, as N_2, is relatively quite abundant in the earth's atmosphere, accounting for 78% by weight of the composition of dry air. However, all animals and most plants are unable to utilize N in that form. The most effective, directly utilizable forms of N for most plants are the nitrate (NO_3^-) and ammonium (NH_4^+) ions. For the large majority of species, nitrate ion is most effective, but some species, particularly under neutral to alkaline conditions, respond better to NH_4^+ than NO_3^-.

There are basically four known processes by which atmospheric N_2 is converted to forms which can readily be absorbed and utilized by plants. One is atmospheric fixation, according to which N_2 reacts with O_2, with catalysis by lightning or ultraviolet radiation, to form oxides (e.g. NO_3^-) which are carried to the soil with rain. This process is believed to account for the addition of no more than 4 to 6 pounds of fixed N per acre per year, which amount just about offsets the losses of fixed N available to plants due to leaching from the upper horizons of the soil.

A process of industrial fixation has been devised, according to which N_2 and H_2 are reacted in the presence of a catalyst at a temperature of about 500° C and several hundred atmospheres pressure to yield ammonia. However, in terms of total N-fixation which occurs in the biosphere, industrial fixation, of course, is a relatively small-magnitude process. Likewise, volcanic expulsion of NH_3 accompanying combustion of igneous rock is a very minor process.

By far the most important process by which atmospheric N_2 is converted to forms utilizable by plants is performed by certain kinds of plant organisms and is therefore known as biological N-fixation. Chemically, biological N-fixation consists of the reduction of N_2 to ammonia:

$$N \equiv N \xrightarrow[\text{Nitrogenase}]{+6e^- + 6H^+} 2\ NH_3$$

Protons and electrons for the endergonic process are generated in metabolic reactions, and catalysis is performed by an enzyme system called nitrogenase.

Many kinds of plants perform this process, either independently or in symbiotic association with other kinds of plants, as the following classification shows.

Table 25-I. Classification of N-fixing Organisms

Type(s) of organism	Example(s)

I. Free-living bacteria
 A. Heterotrophs
    1. Aerobic
    2. Anaerobic
 B. Autotrophs
	
	Azotobacter vinelandii
	Clostridium pasteurianum
	Rhodospirillum rubrum

II. Blue-green algae A. Free-living B. Symbiotic	 Anabaena spp., Nostoc spp. In various lichens and other symbiotic associations, e.g. the fern Azolla and an Anabaena endophyte
III. Bacteria in root nodules of angiosperms A. Legume hosts B. Non-legume hosts	 Soybean (Glycine max) + Rhizobium japonicum Alder (Alnus rubrum), Ceanothus velutinus, and Myrica gale + unidentified microbial symbionts

It is estimated that a vigorous crop of nodulated legumes can fix 200 or more pounds of N per acre per year.

Symbiotic N-fixation in nodules of legumes, with which this exercise is directly concerned, normally is truly a symbiotic function since, in nature, neither the legume host nor the Rhizobium symbiont evidently can fix nitrogen apart from the other. Of course the legume can live and function very well apart from symbiotic association with Rhizobium, obtaining its N by absorbing chiefly nitrate from the soil. The same species of Rhizobium found in nodules also live very well as free-living saprophytes in the soil; however, unlike certain other free-living species of bacteria (e.g. Clostridium and Azotobacter spp.), free-living Rhizobia in nature evidently do not fix nitrogen. Although N-fixation in nodules of legumes strictly depends upon the symbiotic association of the two species, in fact it is the transformed Rhizobium "bacteroids" which perform the actual biochemistry of reducing N_2 to ammonia. This has been determined conclusively by the relatively recent finding that Rhizobia alone can fix nitrogen in vitro under certain laboratory conditions.

A highly significant advance was made independently in several laboratories in about 1966, when it was discovered that the nitrogen-fixing enzyme system, nitrogenase, from Azotobacter vinelandii, Clostridium pasteurianum and soybean nodules (Glycine max + Rhizobium japonicum) is

292

not exclusively specific for N_2 but is capable of catalyzing reduction of several other compounds as well, including reduction of acetylene to ethylene. This discovery has greatly expedited subsequent research on and analysis of N-fixation. Formerly it was necessary to assay N-fixation using the heavy isotope ^{15}N, the analysis of which required mass spectrometry, or some other relatively insensitive, time-consuming procedure. Now N-fixation can be easily assayed at low cost in comparatively little time by measuring the production of ethylene from acetylene using a gas chromatograph. The method is applicable to suspensions of free-living N-fixing bacteria and extracts thereof, as well as whole nodulated plants, nodules and bacteroids isolated from nodules, and cell-free extracts. (Of course, since only two electrons are required to reduce a molecule of acetylene to ethylene, whereas the reduction of a molecule of N_2 to NH_3 requires six electrons, acetylene-reduction data must be divided by a factor of three when converting to equivalent N_2-reduction.)

In this experiment a comparison will be made of the growth of nodulated and non-nodulated soybean plants cultured hydroponically. Nitrogen-fixation will be assayed by the acetylene-reduction method, using whole nodulated soybean plants, excised nodulated root systems and excised nodules.

Materials and Methods[1]

Plant approximately 10 soybean (Glycine max) seeds in each of 6 8-inch plastic pots filled with Perlite. Moisten about 60 additional seeds with distilled water and mix them with Nitragin, a commercial inoculum containing the bacterium Rhizobium japonicum, until the seeds are well-coated with the material. Then plant approximately 10 inoculated seeds in each of 6 additional pots of Perlite. Grow the plants in a greenhouse or growth chamber under a 16-hour photoperiod at approximately 24° C and a light intensity of about 500 to 1,000 ft-c and an 8-hour dark period at approximately 18° C. Irrigate 3 non-inoculated pots and 3 inoculated pots daily with a complete mineral nutrient solution, and irrigate the other 6 pots with a minus-N nutrient

---

[1]Contributed by Dr. Kathleen A. Fishbeck and Dr. Harold J. Evans, both formerly of the Department of Botany and Plant Pathology, Oregon State University, Corvallis; now, respectively, Research Associate in the Department of Agronomy and Plant Genetics, University of Minnesota, St. Paul and Director, Laboratory for Nitrogen Fixation Research, Oregon State University.

solution.[2] Every fourth day irrigate all 12 pots with distilled water, if available, or tap water to prevent salt accumulation.

Approximately 3 to 4 weeks after planting harvest all the plants in each of the 4 groups, except one pot of inoculated plants irrigated with minus-N solution. Measure for each group of plants the: (1) number (percentage) of plants which are nodulated; (2) average shoot length; (3) average fresh weight per plant; and (4) average fresh weight of nodules per plant. Record all harvest data in the table provided in the Report forms.

The single remaining pot of inoculated plants irrigated with minus-N solution is to be tested for their capacity for symbiotic nitrogen fixation. This part of the experiment will be conducted largely as a demonstration, but student participation will be expected.

Place the pot of plants in a polyethylene container which is large enough to accommodate the plants without injury and which can be sealed

---

[2]Composition of minus-N solution:[3,4]

Salt	Concentration (g/liter)
$CaSO_4 \cdot 2 H_2O$	1.033
$MgSO_4 \cdot 7 H_2O$	0.493
K_2SO_4	0.279
KH_2PO_4	0.023
K_2HPO_4	0.145
$CaCl_2$	0.056

Salt (trace elements)	Concentration (mg/liter)
FeEDDHA[5]	16.67
H_3BO_3	1.43
$MnSO_4 \cdot 4 H_2O$	1.02
$ZnSO_4 \cdot 7 H_2O$	0.22
$CuSO_4 \cdot 5 H_2O$	0.08
$CoCl_2 \cdot 4 H_2O$	0.10
$Na_2MoO_4 \cdot 2 H_2O$	0.05

[3]Formulation provided by Harold J. Evans.

[4]Add 105 mg of NH_4NO_3 per liter of solution to make complete nutrient solution.

[5]FeEDDHA is ferric ethylenediamine di-0-hydroxyphenyl acetate, which contains 6% Fe; available from Geigy Chemical Company, P. O. Box 430, Yonkers, New York 10700.

airtight.[6] Seal the container with anhydrous lanolin, silicone grease, or other suitable lubricant.

Calculate the volume of the atmosphere in the container and the volume of acetylene which must be injected to provide a concentration of 1,000 μmoles acetylene/liter. (This is equivalent to 25-cc acetylene per liter and a partial pressure 0.025 atm acetylene.) With a 50-cc syringe withdraw from the sealed container a volume of air equivalent to the calculated volume of acetylene to be injected. Fill the syringe with acetylene from a cylinder of compressed gas and inject the required amount into the container through the serum cap. (Caution: Acetylene is a highly explosive gas; do not expose to sparks or flames, and follow very carefully the directions given by the instructor for filling the syringe from the cylinder!) Mix the atmosphere in the sealed container several times during a succeeding 30-minute incubation period by repeatedly and rapidly filling and emptying a 50-cc syringe through the serum cap. At the end of the incubation period, again mix the atmosphere in the container, and fill 3 1-cc syringes with the atmosphere of the container.

Observe the analysis of the 3 gas samples for ethylene by gas-liquid chromatography, which will be demonstrated by the instructor.[7]

Following analysis of the 3 initial gas samples, open the container, remove the pot of plants, and very carefully excise the root system from each plant. Place all the excised whole root systems back in the container, seal, and repeat the procedure described for the whole plants.

Finally, again open the container, remove the root systems, and harvest the nodules. Put the collective batch of nodules back into the container and repeat the procedure for the acetylene-reduction assay. Present the collective gas chromatography data in the table provided in the Report forms.

To calculate the amount of ethylene produced by the whole plants, excised roots or excised nodules, the following procedure should be followed. First, compare the average height of the ethylene peaks on the strip chart tracings for a group of three replicate samples with a standard curve, which was prepared previously with known amounts of ethylene.

---

[6]The author uses a colored polyethylene container (household utility type) of approximately 21.7-liter capacity and a black plexiglass lid perforated and fitted with a rubber serum cap (No. 2330, Arthur Thomas Company, Philadelphia, Pennsylvania).

[7]The author uses a Varian Aerograph Model 600-D gas chromatograph equipped with a hydrogen flame ionization detector and a Honeywell recorder. The Poropak column (Type R, 80-100 mesh, 30-10 PAK, Waters Associates, Incorporated, Framingham, Massachusetts) is 5 feet long x 1/4 inch in diameter. The oven temperature is 45° C. Nitrogen (N_2) is used as the carrier gas at a flow rate of 75 ml/minute. Under these conditions the retention times for ethylene and acetylene are approximately 58 and 78 seconds, respectively.

By interpolation from the standard curve, the average concentration of ethylene in the group of three 1-cc gas samples is obtained, that is, μmoles C_2H_4/cc. Then multiplication of this concentration by the volume of the atmosphere in the incubation container yields the average number of micromoles of ethylene produced by the plants or excised parts. The instructor will explain any corrections required in the calculations arising from variation in settings (attenuation and range) on the gas chromatograph.

References

1. Bergersen, F. J. 1966. Some properties of nitrogen-fixing breis from soybean root nodules. Biochim. Biophys. Acta 130: 304-312.

2. Bond, G. 1967. Fixation of nitrogen by higher plants other than legumes. Ann. Rev. Plant Physiol. 18: 107-126.

3. Burris, R. H. 1966. Biological nitrogen fixation. Ann. Rev. Plant Physiol. 17: 155-184.

4. Burris, R. H. 1976. Nitrogen fixation. Pp. 887-908 in: J. Bonner and J. E. Varner, Eds. Plant Biochemistry. 3rd Ed. Academic Press, New York.

5. Carnahan, J. E. and J. E. Castle. 1963. Nitrogen fixation. Ann. Rev. Plant Physiol. 14: 125-126.

6. Dilworth, M. J. 1966. Acetylene reduction by nitrogen-fixing preparations from Clostridium pasteurianum. Biochim. Biophys. Acta 127: 285-294.

7. Epstein, E. 1972. Mineral Nutrition of Plants: Principles and Perspectives. John Wiley and Sons, New York.

8. Gauch, H. G. 1972. Inorganic Plant Nutrition. Dowden, Hutchinson, and Ross, Stroudsburg, Pennsylvania.

9. Goodchild, D. J. and F. J. Bergersen. 1966. Electron microscopy of the infection and subsequent development of soybean nodule cells. J. Bacteriol. 92: 204-213.

10. Hardy, R. W. F., R. D. Holsten, E. K. Jackson, and R. C. Burns. 1968. The acetylene-ethylene assay for N_2 fixation: laboratory and field evaluation. Plant Physiol. 48: 1185-1207.

11. Keller, R. A. 1961. Gas chromatography. Scient. Am. 205 (No. 4, Oct): 58-67.

12. Koch, B. and H. J. Evans. 1966. Reduction of acetylene to ethylene by soybean root nodules. Plant Physiol. 41: 1748-1750.

13. Koch, B., H. J. Evans, and S. B. Russell. 1967a. Reduction of acetylene and nitrogen gas by breis and cell-free extracts of soybean root nodules. Plant Physiol. 42: 466-468.

14. Koch, B., H. J. Evans, and S. Russell. 1967b. Properties of the nitrogenase system in cell-free extracts of bacteroids from soybean nodules. Proc. Nat. Acad. Sci. 58: 1343-1350.

15. Mortenson, L. E. 1961. A simple method for measuring nitrogen fixation by cell-free enzyme preparations of Clostridium pasteurianum. Anal. Biochem. 2: 216-220.

16. Postgate, J. R., Ed. 1971. The Chemistry and Biochemistry of Nitrogen Fixation. Plenum Press, London.

17. Raggio, M. and N. Raggio. 1962. Root nodules. Ann. Rev. Plant Physiol. 13: 109-128.

18. Ray, P. M. 1972. The Living Plant. 2nd Ed. Holt, Rinehart and Winston, New York.

19. Salisbury, F. B. and C. W. Ross. 1978. Plant Physiology. 2nd Ed. Wadsworth Publishing Company, Belmont, California.

20. Schöllhorn, R. and R. H. Burris. Study of intermediates in nitrogen fixation. Federation Proc. 25: 710.

21. Stewart, W. D. P. 1966. Nitrogen Fixation in Plants. Athlone Press, University of London.

22. Webster, G. C. 1959. Nitrogen Metabolism in Plants. Row, Peterson and Company, Evanston, Illinois.

Special Materials and Equipment Required
Per Laboratory Section

(120) Soybean (Glycine max) seeds (e.g. variety Chippewa, available from Northrup King and Company, 1500 Jackson Street N.E., Minneapolis, Minnesota 55413)
(6) 8-inch plastic pots
(Approximately 12 to 15 liters) Perlite planting medium
(1 package) Soybean Inoculant ("S" Culture) (Nitragin Company, Incorporated, 3101 West Custer Avenue, Milwaukee, Wisconsin 53209)
(1) Greenhouse or controlled-environment growth chamber
(Stock supply) Complete nutrient solution, prepared as described in Materials and Methods
(Stock supply) Minus-N nutrient solution, prepared as described in Materials and Methods
(1) Polyethylene container, approximately 20-liter capacity, large enough to hold one pot of plants, with a plexiglass lid perforated and fitted with a rubber serum cap
(1) 50-ml syringe

(3) 1-ml syringes
(1) Cylinder of compressed acetylene
(1) Cylinder of compressed ethylene
(1) Gas chromatograph, equipped as described in Materials and Methods

Recommendations for Scheduling

This exercise is most satisfactorily conducted as a class project in each laboratory section, and the gas chromatography generally must be done as a demonstration. To expedite calculations of the amounts of ethylene produced, a standard curve should be prepared with known amounts of the gas in advance. If a gas chromatograph is not available, the acetylene-reduction assay can be omitted and the remainder of the experiment still will constitute an instructive, quite worthwhile exercise.

Duration: 3 to 4 weeks; approximately 30 to 45 minutes required for planting seeds during first laboratory period, about 30 minutes required once weekly thereafter, and 1 entire period is required to terminate the experiment.

EXERCISE 25

SYMBIOTIC NITROGEN FIXATION IN LEGUME NODULES

REPORT

Name _____ Section _____ Date _____

Results of acetylene-reduction assays using nodulated plants supplied with minus-N solution (Group 1):

Plant part(s) assayed	μmoles Ethylene produced			
	Gas sample 1	Gas sample 2	Gas sample 3	Mean
Whole plant	_____	_____	_____	_____
Excised root system	_____	_____	_____	_____
Excised nodules	_____	_____	_____	_____

Example calculation of amount of ethylene produced:

EXERCISE 25

SYMBIOTIC NITROGEN FIXATION IN LEGUME NODULES

REPORT

Name _____ Section _____ Date _____

<u>Comparative growth and nodulation of soybean plants:</u>

Group	No. plants	Conditions — Inoculated	Conditions — Nutrient solution	% of plants nodulated	Average shoot length (cm)	Average fresh weight/plant (g)	Average fresh weight of nodules/ plant (mg)
1		+	Minus N				
2		+	Complete				
3		–	Minus N				
4		–	Complete				

EXERCISE 25

SYMBIOTIC NITROGEN FIXATION IN LEGUME NODULES

REPORT

Name _____ Section _____ Date _____

Questions

1. How did the growth and general appearance of the non-inoculated soy-bean plants supplied with complete nutrient solution compare with that of the inoculated plants supplied with minus-N solution?

2. Is there any evidence from the experiment that ammonium-N or nitrate-N suppressed nodulation or symbiotic N-fixation? Explain.

3. Compare the amounts of ethylene formed (per plant) by whole plants, excised nodulated root systems, and excised nodules. Present possible explanations for any observed differences.

EXERCISE 25

SYMBIOTIC NITROGEN FIXATION IN LEGUME NODULES

REPORT

Name _____ Section _____ Date _____

4. How specific are the symbiotic associations between <u>Rhizobium</u> species and legume species?

5. Briefly discuss the ecological significance of symbiotic N-fixation involving both legume and non-legume seed plants.

6. What are mycorrhizae?

Exercise 26

Phytochrome Effects in Nyctinastic Leaf Movements

Introduction

Autonomous movements of plants have been studied for more than a century. Among the early investigators who were interested in the subject was Charles Darwin, who in 1880 published a remarkable record of observations in a book entitled The Power of Movement in Plants.

Basically, plant movements can be classified as either spontaneous or induced. The most common spontaneous movement is nutation, which consists of the movement of the tip of a growing stem, tendril, or other organ in such a way as to describe a more or less circular path in space. Nutation is a differential growth movement which is manifested as a spiral shifting of the locus of most active growth within the growing region of an organ.

Comprising induced movements are the well-known tropisms. These are growth responses in which the direction of movement is determined by the direction from which the stimulus comes (e.g., phototropism, geotropism, and thigmotropism). The other kind of induced movements are nastic movements. This type of movement results from differential turgor changes instead of differential growth, and the direction of movement is not related to the direction from which the stimulus comes. There are nastic movements in response to light (photonasty), temperature (thermonasty), and touch (thigmonasty), for examples. The flowers of some kinds of plants exhibit regular opening and closing which are typically photonastic or thermonastic movements. For example, flowers of morning-glory, tulip, crocus, and California poppy are open during the day and closed at night. Contrariwise, in some species which are pollinated by night-flying moths, the flowers are closed during the day and open at night.

A well-known example of thigmonasty is the movements of the leaves and leaflets of the sensitive plant (Mimosa pudica). This tropical plant has double pinnately compound leaves, which change position very rapidly due to localized, differential changes in water content and turgor pressure in cells within structures called pulvini. Pulvini are swollen regions at the base of the main petiole (rachis, primary pulvini), secondary petiole (rachilla, secondary pulvini), and leaflets (pinnule, tertiary pulvini). (A pinna is a subdivision of the compound leaf which is composed of about 12 paired pinnules.) When the plant is stimulated by touch or shock, the pinnules fold upward, the main petiole drops, and the secondary petioles move close together. After stimulation uniform turgor of the cells of each pulvinus is gradually regained, and the leaves resume their former orientation.

The best known nastic movements of leaves are the nyctinastic or "sleep movements." Such movements are very common among many species of weeds and cultivated plants. The leaves of bean, clover and oxalis, for examples, are oriented approximately horizontally during the day, take on a vertical orientation or drooping at night, and do not rise again

303

until the next morning. In other species the leaves bend upward at night instead of drooping. In either case, the movements theoretically are an adaptation to comparatively low night temperatures, exposing less of the leaf surface to heat loss by radiation. Obviously the normal leaf orientation during the day is conducive to maximum light absorption and photosynthesis.

This exercise concerns nyctinastic leaf movements in a legume known as Albizzia julibrissin, the silk tree. Let us now review some of the early investigations on nyctinastic leaf movements on which this exercise is based. J. C. Fondeville, H. A. Borthwick and S. B. Hendricks, working at the USDA's agricultural research center at Beltsville, Maryland, published a report in 1966 on extensive studies of nyctinastic leaf movements in Mimosa pudica, a legume which is closely related to Albizzia julibrissin. Incidentally, it is important to note that there evidently is no relationship between the thigmonastic response and the nyctinastic response of the leaves of Mimosa pudica.

Fondeville et al. used excised pinnae, each consisting of approximately 12 paired pinnules. They observed that the pinnae of young Mimosa pudica plants grown on natural days and nights were open or expanded in light and closed in darkness. Closing resulted from the paired pinnules folding together about tertiary pulvini. The closing response at night was evident within 5 minutes after the beginning of darkness and was fully expressed in about 30 minutes, depending somewhat upon the particular pinna and the previous treatment of the plants.

Excised pinnae behaved like those on whole plants and were used in most of the experiments. Among the key observations were that: (a) if excised pinnae were irradiated with far-red light immediately prior to a dark period, the leaflets remained open for several hours in darkness; and (b) if the brief far-red irradiation were followed by red irradiation, the leaflets began to close within 5 minutes and were fully closed within 30 minutes in darkness. The potentiated control of closing movements could be repeatedly established and reversed by alternating brief irradiations with red and far-red light. The effects of red and far-red light and the clearly evident photoreversibility of the red and far-red irradiations indicated that phytochrome was the photoreceptor involved in the response.[1] Moreover, the characteristics of this phytochrome action led Fondeville et al. to suggest that phytochrome, specifically P_{fr}, might act as the molecular level by somehow regulating membrane permeability.

As noted previously, nyctinastic leaf movements are caused by relative changes in turgor pressure (also cell size) on opposite sides of a pulvinus. In the case of the pinnae of Mimosa pudica, in darkness there is a net movement of water out of the cells in the upper side of each tertiary pulvinus into cells of the lower side, causing a folding or closing movement of the leaflets or pinnules. In the light this process is reversed. The transfer of water is caused by an apparent alteration

---

[1]See Introduction to Exercise 10 for more discussion of phytochrome.

of membrane permeability and the net transfer of K^+ and Cl^- ions (perhaps also other electrolytes). The cells comprising pulvini contain extra-ordinarily high concentrations of K^+, on the order of 0.5 M. Leaflet closure is associated with a loss of K^+ and Cl^- from the upper cells and a concomitant loss of water. The lower cells evidently do not gain the ions lost from the upper cells, but the lower cells, because of their negative water potential, do gain water lost from the upper cells. The role of phytochrome in all of this is either to cause a differential transient increase in the permeability of the plasma membrane of the upper cells to K^+ and Cl^-, or perhaps to activate a membrane-bound ATPase.

Quickly the results reported for Mimosa pudica were confirmed and extended by Hillman and Koukkari (1967) and Jaffe and Galston (1967) with Albizzia julibrissin, or silk tree, and some other legumes. A. julibrissin generally is more amenable to investigations of nyctinastic leaf movements than M. pudica because A. julibrissin does not display thigmonasty; it is insensitive to mechanical stimulation. The studies with A. julibrissin readily confirmed that nyctinastic leaf movements in this plant also are under the control of phytochrome, with red light potentiating and far-red light inhibiting pinnule closure in the dark.

Altogether, what emerges regarding nyctinastic leaf movements is the concept that the diurnal change in leaf orientation is controlled by an interaction between an endogenous circadian rhythm and the phytochrome system. This interaction is indicated by the finding that phytochrome control is pronounced only in experiments conducted early in the photo-period and appears to be absent or nearly so toward the end of the light period. A phytochrome-independent, high-intensity light effect actually is responsible for keeping the pinnules open during the photoperiod and for the opening reaction of the pinnules following a dark period.

Materials and Methods[2]

Examine carefully the potted Albizzia julibrissin (silk tree)[3] plants and become familiar with the morphology of the double pinnately compound leaves. Note that there is a rachis or primary petiole to each twice pinnately compound leaf with a primary pulvinus at its base. Paired secondary petioles or rachillae arise from the rachis, and each rachilla bears approximately 12 paired pinnules with a tertiary pulvinus at the base of each. A pinna consists of a rachilla and its paired pinnules.

---

[2]Adopted from Hillman, W. S. and W. L. Koukkari. 1967. Phytochrome effects in the nyctinastic leaf movements of Albizzia julibrissin and some other legumes. Plant Physiol. 42: 1413-1418.

[3]Many other species can be substituted quite successfully. The author has used Acacia cornigera and Oxalis regnellii. Hillman and Koukkari (1967) reported good results with Albizzia lophantha, Leucaena glauca, Poinciana gilliesi and Calliandra inequilatera. The early work by Fondeville et al. (1966) was with the sensitive plant, Mimosa pudica.

The experimental material will be excised pinnule-pairs, each consisting of 2 opposite pinnules connected to a segment of rachilla. Observations and measurements will be made on the opening and closing of the pinnules comprising each pair.

Ideally the plants will have been maintained in a warm (about 24° C) greenhouse under a regimen of 16-hour photoperiods and 8-hour dark periods for at least three days prior to the experiment,[4] and the experiment should be conducted as early as possible after the beginning of a photo-period.[5]

All operations after the start of an experimental dark period should be conducted in a darkroom with minimal exposure of the plants to a dim blue safelight (see Exercise 10). Otherwise, the plant material can be prepared under ordinary laboratory lighting and the excised pinnule pairs floated on water in petri dishes and placed in light-tight boxes that can be fitted with red and far-red cellophane or plastic filters and light-tight covers.

Work in teams of 3 or 4. Select 2 or 3 relatively mature and healthy looking pinnae, enough to yield 49 uniform pinnule pairs, from the potted plants. Pipet 10 ml of distilled water into each of 7 petri dishes. Excise the pinnule pairs with a razor blade or scalpel. Then immediately float 7 pinnule pairs in each petri dish. Number the dishes 1 through 7. In a darkroom or using light-tight boxes equipped with filters, perform the following irradiation treatments,[6] one for each dish:

1. Dark control
2. Red light
3. Far-red light
4. Red light + far-red light
5. Red light + far-red light + red light
6. Red light + far-red light + red light + far-red light
7. White light control

Leave dish no. 7 exposed continuously to laboratory lighting, but transfer all other dishes to darkness immediately after irradiation. Dish no. 1 should be placed in darkness with no prior treatment with either red or far-red light. Parenthetically it can be noted that very brief exposures

---

[4]This is very important; in photoperiods of less than 12 hours the plants go dormant in about two weeks.

[5]This also is very important; the phytochrome effect, hence plant responsiveness, decreases during the course of each photoperiod. Late in the light period the difference between red and far-red treated material is absent or very small. This may reflect an interaction of the phytochrome system with an endogenous circadian rhythm.

[6]Use exposure times of 2 minutes of red light and 4 minutes of far-red light in all the irradiations. Sequential irradiations must follow each other immediately. See Exercise 10 for descriptions of red and far-red light sources and filters.

to room light do not seriously affect the outcome of the treatments. Thirty to 60 minutes after transfer to darkness of dishes 1-6, examine the pinnule pairs in each of the six groups. Take measurements on the pinnule pairs by either: (a) measuring with a protractor the angle of each pinnule axis with the rachilla to which it is attached; or (b) measuring in millimeters the distance between the tips of the pinnules comprising each pair. Record the data in a table.

References

1. Burkholder, P. R. and R. Pratt. 1936. Leaf movement of Mimosa pudica in relation to the intensity and wave length of the incident radiation. Am. J. Botany 23: 46-52 and 212-220.

2. Darwin, C. 1880. The Power of Movement in Plants. D. Appleton and Company, New York.

3. Fondeville, J. C., H. A. Borthwick, and S. B. Hendricks. 1966. Leaflet movement of Mimosa pudica L. indicative of phyto-chrome action. Planta 69: 357-364.

4. Fondeville, J. C., M. J. Schneider, H. A. Borthwick, and S. B. Hendricks. 1967. Photocontrol of Mimosa pudica L. leaf movement. Planta 75: 228-238.

5. Galston, A. W. and R. L. Satter. 1972. A study of the mechanism of phytochrome action. Pp. 51-79 in: V. C. Runeckles and T. C. Tso, Eds. Structural and Functional Aspects of Phytochemistry. Vol. V. Academic Press, New York.

6. Galston, A. W. and R. L. Satter. 1976. Light, clocks, and ion flux: an analysis of leaf movement. Pp. 159-184 in: H. Smith, Ed. Light and Plant Development. Butterworth & Company, Limited, London.

7. Hillman, W. S. 1976. Biological rhythms and physiological timing. Ann. Rev. Plant Physiol. 27: 159-179.

8. Hillman, W. S. and W. L. Koukkari. 1967. Phytochrome effects in the nyctinastic leaf movements of Albizzia julibrissin and some other legumes. Plant Physiol. 42: 1413-1418.

9. Jaffe, M. J. and A. W. Galston. 1967. Phytochrome control of rapid nyctinastic movements and membrane permeability in Albizzia julibrissin. Planta 77: 135-141.

10. Koukkari, W. L. and W. S. Hillman. 1968. Pulvini as the photo-receptors in the phytochrome effect on nyctinasty in Albizzia julibrissin. Plant Physiol. 43: 698-704.

11. Mitrakos, K. and W. Shropshire, Jr., Eds. 1972. Phytochrome. Academic Press, New York.

307

12. Moore, T. C. 1979. Biochemistry and Physiology of Plant Hormones. Springer-Verlag, New York.

13. Palmer, J. H. and G. F. Asprey. 1958. Studies in the nyctin-astic movement of the leaf pinnae of _Samanea saman_ (Jacq.) Merrill. I. A general description of the effect of light on the nyctinastic rhythm. Planta 51: 757-769.

14. Satter, R. L. and A. W. Galston. 1971a. Potassium flux: a common feature of _Albizzia_ leaflet movement controlled by phytochrome or endogenous rhythm. Science 174: 518-520.

15. Satter, R. L. and A. W. Galston. 1971b. Phytochrome-controlled nyctinasty in _Albizzia julibrissin_. III. Interactions between an endogenous rhythm and phytochrome in control of K flux and leaflet movement. Plant Physiol. 48: 740-746.

16. Satter, R. L., P. Marinoff, and A. W. Galston. 1970. Phyto-chrome controlled nyctinasty in _Albizzia julibrissin_. II. Potassium flux as a basis for leaflet movement. Am. J. Botany 57: 916-926.

17. Satter, R. L., D. D. Sabnis, and A. W. Galston. 1970. Phyto-chrome controlled nyctinasty in _Albizzia julibrissin_. I. Anatomy and fine structure of the pulvinule. Am. J. Botany 57: 374-381.

Special Materials and Equipment Required
Per Team of 3-4 Students

Seeds as well as potted plants of _Albizzia julibrissin_ are available commercially. The seeds have heavy coats, hence scarification with an abrasive solid material or with concentrated sulfuric acid aids germina-tion. Plants must be maintained under relatively long days (16 hours) and short nights (8 hours); they will go dormant within about two weeks if kept in photoperiods of 12 hours or less.

(1) Potted _Albizzia julibrissin_ plant with several mature, healthy leaves
(1) Irradiation apparatus (if a darkroom is to be used) (see Exercise 10 for description)
(1) Darkroom
(1) Set of light-tight boxes that can be covered, first, with red or far-red cellophane or plastic filters, and second, with light-tight covers
(7) 10-cm petri dishes
(1) Protractor or metric ruler

Recommendations for Scheduling

Potted plants should be available at the beginning of the laboratory period when the experiment is to be conducted. Since the results of the irradiations with red and far-red light vary markedly with time after beginning of the photoperiod, the experiment should be performed <u>as soon as possible after the beginning of a photoperiod</u>. For laboratory periods scheduled in the late morning and afternoon, it is necessary to hold the plants under artificial regimens of light (16 hours) and dark (8 hours) for several days or longer before the date on which the experiment is performed.

Students should work in teams of 3 to 4 members. If a darkroom is used during the irradiation treatments, one or two students each from all teams can go to the darkroom at one time. Alternatively, whole teams can individually take turns using the darkroom.

<u>Duration</u>: One 2- to 3-hour laboratory period; nearly exclusive attention to this exercise required.

PHYTOCHROME EFFECTS IN NYCTINASTIC LEAF MOVEMENTS

REPORT

Name _____ Section _____ Date _____

Measurements of pinnule opening and closing:

Light treatment	Distance between tips of pinnules comprising each pair (mm)[1]							
	Pair 1	Pair 2	Pair 3	Pair 4	Pair 5	Pair 6	Pair 7	Mean
1. Dark control								
2. R								
3. FR								
4. R + FR								
5. R + FR + R								
6. R + FR + R + FR								
7. White light control								

[1]Alternatively, angles of pinnule axes with rachillas (°).

PHYTOCHROME EFFECTS IN NYCTINASTIC LEAF MOVEMENTS

REPORT

Name _____ Section _____ Date _____

Questions

1. What is the specific evidence, if any, from this experiment that
    phytochrome is active as a photoreceptor in nyctinastic leaf
    movements of <u>Albizzia</u>?

2. Describe the nature of the evidence from other investigations that an
    endogenous circadian rhythm interacts with the phytochrome system in
    controlling nyctinastic leaf movements in plants like <u>Albizzia</u>.

PHYTOCHROME EFFECTS IN NYCTINASTIC LEAF MOVEMENTS

REPORT

Name _____ Section _____ Date _____

3. Of what possible adaptive or survival value are nyctinastic leaf movements to those plants that display them?

4. Identify some of the most important sources or causes of variation in an experiment like the one you performed.

5. Why was the exposure time for far-red irradiation twice that for red irradiation in this experiment?

PHYTOCHROME EFFECTS IN NYCTINASTIC LEAF MOVEMENTS

REPORT

Name _____ Section _____ Date _____

6. Describe the nature of the changing ratio of P_{fr}/P_r that occurs in intact leaves of <u>Albizzia</u> during a normal photoperiodic cycle consisting of a 16-hour day and an 8-hour night, and relate how phytochrome participates in the control of the nyctinastic leaf movements that occur under these conditions.

MEASUREMENT OF LEAF WATER POTENTIALS
WITH A PRESSURE CHAMBER

Introduction

The cohesion theory, first proposed by Dixon and Joly in 1894, has long been recognized as the only consistent theory to explain how xylem sap can be raised to the tops of the tallest trees. According to this theory, the ramified water column in a plant is continuous from the root epidermis to the mesophyll in the leaves. During transpiration, water evaporates from the mesophyll and leaves the plant via the stomata. As water evaporates from the leaf or is used for other processes in the leaf mesophyll, a suction or negative pressure is exerted upon the water in the complex system of conducting elements of the xylem. This negative pressure is transmitted through the plant to the root, where it induces water to move into the plant from the soil. The water column does not break because of the high tensile strength of water and because of the small diameter of the conducting elements which prevents cavitation. A measure of the negative pressure is a direct indication of the important physiological parameter which is called variously, "sap pressure" (meaning water potential of the sylem sap), "leaf water potential," "plant water potential," or "plant moisture stress." Most commonly it is measured in bars (1 bar = 0.987 atm); it varies from zero to sometimes quite large negative values.

Until relatively recently, rapid and reliable methods for measuring leaf water potential were unavailable. Then in 1965 Scholander et al. described a pressure chamber technique, which was first introduced by Dixon (1914). This is a very rapid, simple and reliable technique which readily can be utilized in the field as well as the laboratory.

The theory on which the pressure chamber technique is based is relatively quite simple. Under equilibrium conditions, that is when water loss is zero, the water potential of leaf cells is equal to the water potential of xylem sap. Within the xylem the water column is generally under tension. This tension results from the demands of the leaves for replacement of water lost to the atmosphere and from the inability of the roots to absorb water rapidly enough to alleviate the stress. The water potential of the xylem sap is accurately described by the expression:

$$\Psi_w = \Psi_s + \Psi_p$$

where Ψ_w = water potential
        Ψ_s = solute potential (sometimes designated as Ψ_π)
        Ψ_p = pressure potential

However, in most cases the effect of solutes on the potential of the xylem sap (Ψ_s) is relatively small (e.g., -0.5 to -1.0 bar), and Ψ_s is ignored. Thus $\Psi_w \simeq \Psi_p$. The Ψ_p, negative because the xylem sap is under tension, can be measured with a pressure chamber. That is:

$$\Psi_w \simeq \Psi_p = -P$$

where P = the external pressure required in a
pressure chamber to equalize Ψ_w
(sometimes called "plant moisture
stress")

Since under equilibrium leaf water potential equals the water potential
of the xylem sap, P is a measure, equally well, of either leaf water
potential or xylem sap potential. When the transpiration rate is not
zero, however, leaf water potential is invariably somewhat lower than
the xylem water potential.

In practice, the pressure chamber technique is a simple procedure.
When a leaf or twig is severed from a plant shoot, the water column in
the xylem is broken and, because it is under tension, the xylem sap pulls
back or recedes into the twig or leaf a short distance. To measure the
original tension (that is, to measure P), the twig or leaf is sealed in
a chamber with only the cut end exposed to atmospheric pressure. The
pressure in the chamber is then increased until xylem sap is forced back
to the protruding cut surface. The pressure required to force xylem
sap back to the cut surface is equal in magnitude but opposite in sign
to the tension (negative pressure) the xylem sap was under originally.
If a relatively large amount of pressure is necessary, the plant is
said to have a relatively low leaf water potential and a relatively
high moisture stress.

Materials and Methods[1]

This exercise will be performed as a class demonstration, with
direct participation by as many students as possible.

A. Becoming familiar with apparatus. First, become familiar with
the pressure chamber apparatus[2] and its safe operation by reading the entire Materials and Methods section. Practice taking
measurements of leaf water potential using the conifer seedlings provided. The following steps will ensure safe and
effective operation of the apparatus:

1. Cut a whole plant shoot or twig about 12 cm below the terminal bud. Use a sharp razor blade or scalpel and cut
diagonally to produce a smooth, clean surface. In woody
species, the bark must be stripped back far enough to allow

---

[1]Based in part on: Cleary, B. D. and J. B. Zaerr. 1980. Pressure
chamber techniques for monitoring and evaluating seedling water status.
New Zealand J. For. Sci. 10: 133-141.

[2]Pressure chambers are available from: PMS Instrument Company,
2750 N. W. Royal Oaks Dirve, Corvallis, OR 97330; and Soil Moisture
Equipment Corporation, P. O. Box 30025, Santa Barbara, CA 93105.

the woody portion to protrude through the rubber sealing gasket. Failure to remove the bark may result in mistaking phloem exudate for xylem sap at the endpoint. Resin can also obscure the endpoint in species with resin tubes, particularly the pines, but resin usually forms bubbles which break and can be wiped away.

2. Insert the cut end of the twig through the hole in the rubber gasket. If the cut surface of the twig is unclean or damaged, use a razor blade to remove another thin layer from the cut end. Approximately 90% of the sample should be mounted so as to be inside the pressure chamber. Avoid kinking the twig. Insert tender twigs or shoot cuttings with a convenient insertion tool such as a narrow probe. Install the mounted twig in the chamber. If more than two minutes elapse between the time the twig is cut and the time it is sealed in the chamber, discard the twig and start again with a fresh one.

3. Increase pressure in the chamber <u>slowly</u> while observing the cut surface of the twig. When water first appears on the cut surface, record the chamber pressure. Use of reflected light and a magnifying hand lens will expedite observing the exuded water. Do not repeat measurements on the same tissue sample.

4. Exhaust the chamber <u>slowly</u>; remove the twig and prepare the next sample.

NOTE: <u>IMPORTANT SAFETY PRECAUTIONS</u>

The pressure chamber apparatus is a potentially dangerous device which should be treated with great respect. It can be extremely useful but also can injure a careless operator. Because the device uses compressed inert gas (preferably nitrogen) as its pressure source, failure of the pressurized vessel may cause the sudden release of pressure similar to an explosion. Sudden failure of the connecting hose or tubing, though less serious, could result in a wildly thrashing hose or, possibly, flying parts.

To safely operate the pressure chamber, establish and use the following routine for applying pressure to the instrument:

1. Turn the instrument's control valve to "off."

2. Check the supply hose to see that it is connected properly. Check, by careful inspection and tugging, hose fittings and any other connections that might be loose.

3. Open the valve on the supply tank <u>slowly</u>, not suddenly. If the system should fail anywhere, the more nearly closed this valve is, the less damage is likely to result. This same principle--open the valve slowly--also applies to filling a portable tank from a main supply tank.

4. Do not turn the valve wide open! Instead, open it only wide enough to permit the chamber to be pressurized without a substantial drop in hose pressure. One-half turn is usually sufficient. Close the main tank valve at the end of the day to conserve gas in case a small leak develops.

5. Before making any measurements, seal the chamber by inserting a solid rubber stopper in the lid in place of a tissue sample, allow compressed gas to enter the chamber, and adjust the rate control valve so that pressure increases at a rate of approximately 10 psi/sec (or 0.7 bars/sec).

6. Test the instrument's safety equipment to help detect leaks and perform minor adjustments on valves. Clean and lubricate chamber seals because dirt and foliage cause wear and can even plug valves or tubing.

Following these safety guidelines will not only minimize the inherent hazards of the technique but will also improve the efficiency of collecting data.

During operation:

1. Never place any part of the body--particularly the eyes-- directly above the hole in the chamber cover. Because fairly high pressures are required to force a plant through the seal, such an event can occur unexpectedly and with considerable vigor. The endpoint can be observed just as easily from the side as from above.

2. Never leave the chamber cover or lid half on. It should always be either on and ready to be pressurized or completely off. A cover setting on the chamber but not locked in place can form enough of a seal to withstand low pressure, but may blow off at higher pressures. Avoid this situation by always leaving the chamber cover all the way on or all the way off.

3. After completing the day's measurements, close the main tank valve and leave the control valve on off or exhaust. Again, leave the cover all the way on or all the way off. If removing the supply hose is necessary, ascertain that the hose is not pressurized before disconnecting it. Like the chamber cover, quick-disconnect fittings should be either completely connected or completely disconnected. Fittings should be disconnected only when the hose is not pressurized.

B. Experimental. Select one large conifer or other woody plant with suitable branches or ten uniform, well-watered conifer or other woody plants which have been kept under high humidity and low light intensity or darkness since early morning. Measure the leaf water potential (Ψ_w) of two plants (or two branches on a large plant) selected at random. Place all

318

remaining plants under bright lights (Time 0) and measure the Ψ_w at 10-minute intervals thereafter for 1.5 to 2.0 hours. Plot the results on the graph paper provided. If two or more species are available, compare their responses. As a variation, do a similar experiment with wind, or a combination of light and wind.

References

1. Barris, H. D. 1965. Comparison of water potentials in leaves as measured by two types of thermocouple psychrometer. Australian J. Biol. Sci. 18: 36-52.

2. Boyer, J. S. 1966. Isopiestic technique: measurement of accurate leaf water potentials. Science 154: 1459-1460.

3. Boyer, J. S. 1967. Leaf water potentials measured with a pressure chamber. Plant Physiol. 42: 133-137.

4. Boyer, J. S. and E. B. Knipling. 1965. Isopiestic technique for measuring leaf water potentials with a thermocouple psychrometer. Proc. Nat. Acad. Sci. 54: 1044-1051.

5. Briggs, G. E. 1967. Movement of Water in Plants. Davis Publishing Company, Philadelphia.

6. Cleary, B. D. and J. B. Zaerr. 1980. Pressure chamber techniques for monitoring and evaluating seedling water status. New Zealand J. For. Sci. 10: 133-141.

7. Dixon, H. H. 1914. Transpiration and the Ascent of Sap in Plants. Macmillan Company, New York.

8. Dixon, H. H. and J. Joly. 1894. On the ascent of sap. Ann. Bot. 8: 468-470.

9. Hinckley, T. M., J. P. Lassoie, and S. W. Running. 1978. Temperal and spatial variations in the water status of forest trees. Forest Science Monograph 20. 72 Pp.

10. Kaufmann, M. R. 1968. Evaluation of the pressure chamber technique for estimating plant water potential of forest tree species. For. Sci. 14: 369-374.

11. Knipling, E. B. 1967. Measurement of leaf water potential by the dye method. Ecology 48: 1038-1041.

12. Knipling, E. B. and P. J. Kramer. 1967. Comparison of the dye method with the thermocouple psychrometer for measuring leaf water potentials. Plant Physiol. 42: 1315-1320.

13. Kramer, P. J. 1969. Plant and Soil Water Relationships: A Modern Synthesis. McGraw-Hill Book Company, New York.

14. Kramer, P. J., E. B. Knipling, and L. N. Miller, 1966. Terminology of cell-water relations. Science 153: 889-890.

15. Ritchie, G. A. and T. M. Hinckley. 1975. The pressure chamber as an instrument for ecological research. Adv. Ecol. Res. 9: 165-254.

16. Salisbury, F. B. and C. W. Ross. 1978. Plant Physiology. 2nd Ed. Wadsworth Publishing Company, Belmont, California.

17. Scholander, P. F., E. D. Bradstreet, H. T. Hammel, and E. A. Hemmingsen. 1966. Sap concentrations in halophytes and some other plants. Plant Physiol. 41: 529-532.

18. Scholander, P. F., H. T. Hammel, E. D. Bradstreet, and E. A. Hemmingsen. 1965. Sap pressure in vascular plants. Science 148: 339-346.

19. Scholander, P. F., H. T. Hammel, E. A. Hemmingsen, and E. A. Bradstreet. 1964. Hydrostatic pressure and osmotic potential in leaves of mangroves and some other plants. Proc. Nat. Acad. Sci. 52: 119-125.

20. Slatyer, R. O. 1967. Plant-Water Relationships. Academic Press, New York.

21. Waring, R. H. and B. D. Cleary. 1967. Plant moisture stress: evaluation by pressure bomb. Science 155: 1248-1254.

Special Materials and Equipment Required
for Laboratory Section

(1) Pressure chamber apparatus
(1) Flat of conifer seedlings or other woody plants with shoot lengths of at least 15 to 20 cm
(2 or 3) Potted conifers or other woody plants extensively branched and with shoot lengths of approximately 0.25 to 0.50 meter

Recommendations for Scheduling

It is recommended that this exercise be conducted essentially as a class demonstration, with participation by as many students as possible.

Duration: One laboratory period.

MEASUREMENT OF LEAF WATER POTENTIALS
WITH A PRESSURE CHAMBER

REPORT

Name _____ Section _____ Date _____

Measurements of leaf-water potentials:

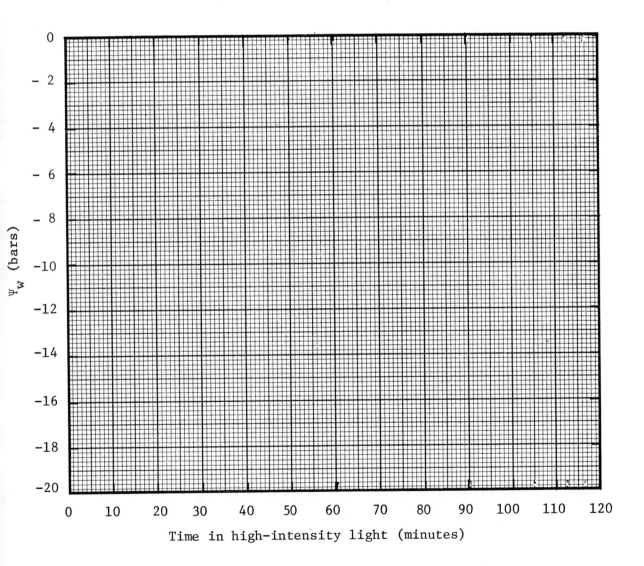

MEASUREMENT OF LEAF WATER POTENTIALS
WITH A PRESSURE CHAMBER

REPORT

Name _____ Section _____ Date _____

Questions

1. Explain why some investigators prefer the term "plant moisture stress" to use of "pressure potential of the xylem sap" or "xylem pressure potential."

2. In what ways does xylem pressure potential differ from water potential of the xylem sap? from leaf water potential?

3. Describe an alternate method of measuring leaf water potential to the pressure chamber technique.

MEASUREMENT OF LEAF WATER POTENTIALS
WITH A PRESSURE CHAMBER

REPORT

Name _____ Section _____ Date _____

4. Sketch a graph illustrating the variation in leaf water potential that a healthy plant growing in a moist soil (at or above field capacity) would be expected to manifest during a typical 24-hour day.

5. Describe briefly the general effects of at least three physical environmental factors on the rate of transpiration in terrestrial seed plants.

6. How is transpiration rate related to leaf water potential?

MEASUREMENT OF LEAF WATER POTENTIALS
WITH A PRESSURE CHAMBER

REPORT

Name _____ Section _____ Date _____

7. What kinds of plants generally develop the most negative (lowest) leaf water potentials? What is the explanation for their unusually low values?

8. In some ring-porous oaks, some large xylem vessels are air-filled in the intact plant. How might this affect estimates of Ψ_{leaf} using the pressure chamber?

THE HILL REACTION OF PHOTOSYNTHESIS

Introduction

Green plants ordinarily do not evolve detectable quantities of oxygen via photosynthesis if they are illuminated in the absence of CO_2. What O_2 is produced is obscured by the rapid uptake of O_2 in respiration. The reactions comprising O_2 evolution are closely coupled to the reactions comprising CO_2 fixation in the intact plant. In the intact leaf, electrons (initially accompanied by H^+) produced by the photolytic cleavage of H_2O are accepted first by acceptor Q, then are passed to a series of electron carriers, and ultimately participate in the reactions of CO_2 fixation.

An important advance in studies of the partial reactions of photosynthesis was made in 1937, when the English biochemist Robin Hill first achieved an experimental separation of O_2 evolution and CO_2 fixation. Hill's technique was to use isolated chloroplasts and chloroplast fragments, and to supply them with a suitable reagent to accept the electrons (and H^+) produced via the photolysis of water. The isolation of chloroplasts by Hill's procedures results in losses of some of the essential endogenous electron acceptors.

Thus, oxygen evolution by illuminated suspensions of chloroplasts can readily be demonstrated using an artificial electron acceptor such as benzoquinone:

or with 2,6-dichlorophenol indophenol (DCPIP), which is blue in the oxidized quinone form and which becomes colorless when reduced to the phenolic form:

$$DCPIP + H_2O \longrightarrow DCPIP\text{-}H_2 + \tfrac{1}{2}O_2$$

In the literature the partial processes of photosynthesis that result in O_2 production are commonly referred to as the "Hill reaction." Hill's pioneering work not only yielded valuable new information directly about the partial reactions of photosynthesis but also led subsequently to other findings about the intricacies of the two photosystems of photosynthesis.

Interestingly, several compounds which are known primarily as herbicides are potent inhibitors of the Hill reaction, and some of these are used to advantage in probing electron transport. One such herbicide is 3-(3,4-dichlorophenyl)-1,1-dimethylurea, also known as diuron, which is used in this exercise. Others are monuron, atrazine, and simazine.

Materials and Methods

The entire class can work as a single group in the preparation of the chloroplasts, with teams of 3 or 4 then performing the remaining parts of the exercise. Alternatively, each team can perform the entire exercise.

Isolation of spinach chloroplasts. Prepare the chloroplast suspension from spinach leaves which were obtained at a local market and kept moist and refrigerated in darkness prior to use. First wash the spinach, discarding obviously unhealthy tissues. Remove the petioles, midribs and larger veins simply by tearing the leaves apart by hand and discarding the veins. Then blot the leaf fragments and, with a sharp blade, cut them into pieces about 1 cm square. Each group will require 30 g of tissue. During these and all subsequent steps in the chloroplast isolation, take care to keep the leaf material, glassware and all solutions on ice.

Now transfer the leaf pieces to a Waring blender, and add 125 ml of cold "isolation medium" containing 0.4 M sucrose, 0.03 M KCl and 0.02 M Tricine buffer adjusted to pH 7.5 with KOH. Blend the mixture for 15 seconds at maximum speed. Scrape off any unbroken tissue from the wall of the blender vessel and blend for an additional 10 seconds. Filter the homogenate through 4 layers of cheesecloth. Collect the filtrate in a chilled beaker and transfer it to centrifuge bottles or tubes. Balance the tubes or bottles precisely (use pipet) against identical vessels containing either filtrate or water.

Centrifuge the filtrate at 500 x g for 5 minutes. Discard the pellet and recentrifuge the supernatant at 2,000 x g for 15 minutes. Now gently suspend the resulting chloroplast pellet in 10 ml of cold "isolation medium" using a glass rod. To break up lumps, if any, pour the suspension through a loosely packed wad of cotton placed on the stem of a short-stemmed funnel. Keep the chloroplast suspension on ice.

Pipet 0.1 ml of the chloroplast suspension into a 25-ml volumetric flask, and make up to volume with 80% (v/v) acetone. Mix well (centrifuge if not clear) and measure the absorbance of the resulting chlorophyll extract in a colorimeter or spectrophotometer at 652 nm. From this measurement calculate the chlorophyll concentration (in mg/ml) in the chlorophyll extract, using the following formula[1] and solving for "c":

$$A = Ecd$$

where A = observed absorbance

E = a proportionality constant known as the absorbancy index (36 ml/cm mg in this case)

---

[1]Alternatively, simply multiply the absorbance by 0.58 to obtain the chlorophyll concentration (in mg/ml) in the stock chloroplast suspension.

c = concentration in mg/ml

d = distance of light path (1 cm in this case)

Record the chlorophyll concentration in the extract. <u>Then correct for the dilution factor, and calculate the chlorophyll concentration in the chloroplast suspension.</u>

<u>Demonstration of the Hill reaction and the effects of chlorophyll concentration, light intensity and DCMU.</u> Prepare 10 ml of a dilute chloroplast suspension containing approximately 0.02 mg/ml chlorophyll by diluting an appropriate aliquot of the stock chloroplast suspension to 10 ml with cold isolation medium. Select 9 test tubes of equal size and prepare the following incubation mixtures for testing the effects of chlorophyll concentration, light intensity and DCMU on the rate of reduction of DCPIP. (<u>Note the instructions that follow</u> on timing of additions of chloroplast suspension.)

Tube no.	0.02 M Tricine buffer (ml)	Dil. chloro. suspension (ml)	Isolation medium (ml)	DCMU solution[2] (ml)	DCPIP solution[3] (ml)
1	3.0	None	1.0	None	1.0
2	4.0	0.5	0.5	None	None
3	4.0	1.0	None	None	None
4 & 5	3.0	0.5	0.5	None	1.0
6 & 7	3.0	1.0	None	None	1.0
8 & 9	None	1.0	None	3.0	1.0

Measure the absorbance of tube no. 1 at 600 nm using a blank consisting of 1.0 ml of "isolation medium" mixed with 4.0 ml of Tricine buffer. Add the indicated amounts of Tricine buffer, isolation medium, DCPIP and DCMU to each of the tubes nos. 2 through 9. Next, add 0.5 ml of the dilute chloroplast suspension (0.02 mg chlorophyll/ml) to tube no. 2 and 1.0 ml of the dilute suspension to tube no. 3. Immediately measure the absorbance of both these tubes at 600 nm, using a blank consisting of 1.0 ml of isolation medium and 4.0 ml of Tricine buffer.

The next step is to place tubes nos. 4, 6 and 8 in a rack positioned 30 cm from a high-intensity light source. A 75- or 100-watt incandescent lamp and a heat filter consisting of a narrow glass container filled with water works well. Place tubes nos. 5, 7 and 9 in a rack positioned 60 cm from the light source. Quickly add 0.5 ml of the dilute chloroplast

---

[2]Stock solution of 5 μM in Tricine buffer.

[3]Stock solution of 0.2 mM in Tricine buffer.

suspension to tubes nos. 4 and 5. Add 1.0 ml of the suspension each to tubes 6, 7, 8 and 9. NOTE THE TIME. When tube no. 6 (highest chlorophyll concentration and highest light intensity) appears to have lost most of its blue color, again note the time. Quickly cover all tubes with aluminum foil and place them on ice. Measure the absorbance of each reaction mixture at 600 nm using the same blank as before. Calculate the initial absorbance of the 6 reaction mixtures (using the data for tubes No. 1, 2 and 3) and the decrease in absorbance after the light treatment. Report the data in a table.

References

1. Arnon, D. I. 1960. The role of light in photosynthesis. Scient. Am. 203: 104-118.

2. Arnon, D. I. 1967. Photosynthetic activity of isolated chloroplasts. Physiol. Rev. 47: 317-358.

3. Audus, L. J., Ed. 1964. The Physiology and Biochemistry of Herbicides. Academic Press, New York.

4. Avron, M. 1975. The electron transport chain in chloroplasts. Pp. 374-386 in: Govindjee, Ed. Bioenergetics of Photosynthesis. Academic Press, New York.

5. Bishop, N. I. 1958. The influence of the herbicide, DCMU, on the oxygen-evolving system of photosynthesis. Biochim. Biophys. Acta 27: 205-206.

6. Bonner, J. and A. W. Galston. 1952. Principles of Plant Physiology. W. H. Freeman and Company, San Francisco.

7. Devlin, R. M. and A. V. Barker. 1971. Photosynthesis. Van Nostrand Reinhold Company, New York.

8. Govindjee and R. Govindjee. 1974. The primary events of photosynthesis. Scient. Am. 231: 68-82.

9. Gregory, R. P. F. 1977. Biochemistry of Photosynthesis. 2nd Ed. John Wiley & Sons, New York.

10. Hill, R. and C. P. Whittingham. 1957. Photosynthesis. 2nd Ed. John Wiley & Sons, New York.

11. Kok, B. 1976. Photosynthesis: the path of energy. Pp. 845-885 in: J. Bonner and J. E. Varner, Eds. Plant Biochemistry. 3rd Ed. Academic Press, New York.

12. Levine, R. P. 1969. The mechanism of photosynthesis. Scient. Am. 221: 58-70.

13. Pfister, K. and C. J. Arntzen. 1979. The mode of action of photosystem II-specific inhibitors in herbicide-resistant weed biotypes. Z. Naturforsch. 34c: 996-1009.

14. Rabinowitch, E. and Govindjee. 1969. Photosynthesis. John Wiley & Sons, New York.

15. Salisbury, F. B. and C. W. Ross. 1978. Plant Physiology. 2nd Ed. Wadsworth Publishing Company, Belmont, California.

16. Simonis, W. and W. Urbach. 1973. Photophosphorylation in vivo. Ann. Rev. Plant Physiol. 24: 89-114.

17. Stemler, A. and R. Radmer. 1975. Source of photosynthetic oxygen in bicarbonate-stimulated Hill reaction. Science 190: 457-458.

18. Wessels, J. S. C. and R. van der Veen. 1956. The action of some derivatives of phenylurethan and of 3-phenyl-1,1-di-methylurea on the Hill reaction. Biochim. Biophys. Acta 19: 548-549.

Special Materials and Equipment Required Per Team

(Approximately 100 g) Fresh spinach, which should be kept moist and refrigerated in darkness
(1) Centrifuge
(1) Balance
(1) Colorimeter or spectrophotometer, with 1 cm cuvettes
(250 ml) "Isolation medium," containing 0.4 M sucrose, 0.03 M KCl and 0.02 M Tricine buffer (pH 7.5)
(100 ml) Tricine buffer (0.02 M, pH 7.5)
(50 ml) 80% Acetone
(10 ml) DCPIP solution (0.2 mM in Tricine buffer)
(10 ml) DCMU solution (5 μM in Tricine buffer)
(2 to 4) Plastic centrifuge bottles or tubes
(Several) Pipets
(1) Blender
(1) Incandescent lamp (75- or 100-watt)
(1) Glass tank
(12) Test tubes
(2) Test tube racks
(1) Volumetric flask (25-ml)

Recommendations for Scheduling

It is recommended that the whole class or 2 or 3 large teams prepare the isolated chloroplasts, and that the remainder of the exercise be done by smaller teams of 3 or 4 students. Stock solutions can be prepared and kept cold prior to the laboratory period.

<u>Duration</u>: One laboratory period; exclusive attention to this exercise required.

EXERCISE 28

THE HILL REACTION OF PHOTOSYNTHESIS

REPORT

Name _____ Section _____ Date _____

Tube no.	Components and conditions	Initial absorbance (calculated)[1]	Measured absorbance	Change in absorbance
1	No chloroplast suspension	--	_____	--
2	0.5 ml chloroplast suspension, no DCPIP	--	_____	--
3	1.0 ml chloroplast suspension, no DCPIP	--	_____	--
4	0.5 ml chloroplast suspension, high-intensity light	_____	_____	_____
5	0.5 ml chloroplast suspension, low-intensity light	_____	_____	_____
6	1.0 ml chloroplast suspension, high intensity light	_____	_____	_____
7	1.0 ml chloroplast suspension, low-intensity light	_____	_____	_____
8	1.0 ml chloroplast suspension, high-intensity light, DCMU	_____	_____	_____
9	1.0 ml chloroplast suspension, low-intensity light, DCMU	_____	_____	_____

[1]For tubes containing 0.5 ml of chloroplast suspension, the initial absorbance is the sum of the measured absorbance of tubes nos. 1 and 2, and for tubes containing 1.0 ml of chloroplast suspension, the initial absorbance is the sum of the measured absorbance of tubes 1 and 3.

EXERCISE 28

THE HILL REACTION OF PHOTOSYNTHESIS

REPORT

Name _____ Section _____ Date _____

Questions

1. Did DCMU inhibit the Hill reaction? If so, explain the observed
    inhibition.

2. Identify one or more herbicides besides DCMU that inhibit photo-
    synthesis via inhibition specifically of the Hill reaction.

3. Would photosynthetic phosphorylation (production of ATP) be expected
    to occur in isolated chloroplasts under any of the conditions
    employed in this exercise? Explain.

THE HILL REACTION OF PHOTOSYNTHESIS

REPORT

Name _____ Section_____ Date _____

4. Describe an alternative method for measuring the Hill reaction.

5. Diagram in as much detail as possible the normal pathway of the flow of electrons from H_2O through Photosystems II and I in a normal, intact leaf, and show where, in this experiment, electrons were accepted instead by DCPIP.

α-amylase, <u>de</u> <u>novo</u> synthesis in barley aleurone cells   175–189
<u>Anabaena</u>   292
anthesin   245
anthocyanin   109
apical control   153
__ dominance   153–161, 215
arseno-molybdate reagent   178
ash trees   121–129
aspartic acid   68
atmospheric nitrogen fixation   291
atrazine   216, 218
<u>Atriplex</u>   68, 255
autoradiogram   279
autoradiography   279
auxin, effect on cucumber hypocotyl elongation   229–241
__, effect on morphogenesis in tobacco callus   95–105
__, general discussion of   96–97, 153–154, 215–216, 229–231
__, naturally-occurring   96, 153–154, 215, 229–231
__, role in apical dominance   153–161
__, structural formula of   96, 154, 230
__, synthetic   215–216, 217
__ transport   204
auxin-type herbicides   216, 217
<u>Avena</u> <u>sativa</u>   267
axillary buds   153–154, 204
<u>Azolla</u>   292
<u>Azotobacter</u>   292

B-995, effects on pea and barley   215–228
__, general discussion of   215, 219, 229–231
__, interaction with auxin   229–241
__, interaction with gibberellin   229–241
__, preparation of solutions of   219
__, structural formula of   217, 229
BA (see benzyladenine)
bacteria, nitrogen-fixing   291–293
__, photosynthetic   35, 292
bacteroids   292
bakanae   163
barley, effects of growth regulators on   215–228
__ endosperm bioassay for gibberellins   175–179
__ grain, longitudinal section of   176
bean   68, 192
beet   6, 8
benzene, effect on membrane permeability   7, 10
benzyladenine, effect on bean leaf growth and senescence   191–202
__, effect on dormancy in <u>Lemna</u> <u>minor</u>   143–151
__, preparation of solutions of   144–145, 194
__, structural formula   144, 192
benzoquinone   325
betacyanin   6
<u>Beta</u> <u>vulgaris</u>   6, 8, 68
<u>Betula</u> <u>pubescens</u>   132
biennials   131, 163

cellular differentiation 81, 95-97
__ organelles 5-6
Cercidiphyllum japonicum 153
chelates, iron 98, 256, 259, 294
chemical potential, of water 12-14, 315-316
Chenopodium rubrum 245
chilling, effect on bud dormancy 132
__, effect on seed dormancy 121-129
chloramphenicol 177
Chlorella 47, 52
chloride ion 83, 255
chlorinated aliphatic acids 216
chlorine 255
(2-chloroethyl)-trimethylammonium chloride (see CCC)
chlorophenoxy herbicides 215-219
chlorophylls, absorption spectra of 38-40
__, extraction from senescent bean leaves 194
__, extraction from spinach 35-45
__, structural formulae of 36
__, thin-layer chromatography of 35-45
chloroplast 5, 11, 35, 67
__ pigments 35-37
chromatogram 37, 43
chromatography 37-40, 293-296
chromoplasts 35
circadian rhythms 245, 305
circulation 277
citrate buffer 25
Clostridium 292
cobalt 255
cocklebur 192, 245, 246, 247
cohesion theory 315
colorimeter 7, 39, 41, 178, 194, 268
colorimetry 7, 39, 178, 194, 268
competitive inhibition (interaction) 230-231
complete mineral nutrient solution 144, 155, 164, 169, 219, 222, 232,
    256, 278, 294
copper 255
corn 68, 70, 74, 219, 222, 245
Corylus avellana 122, 133-134, 136
counting solution 60, 61, 72, 75
counts per minute (cpm) 60, 72, 279
critical daylength 244
__ nightlength 244
Crown Gall 96
cucumber 231, 235
Cucumis sativus 235
culture medium, for Fusarium moniliforme 164, 166
__ __, for Scenedesmus obliquus 47-48
__ __, for tobacco callus 97-98
cycloheximide 177, 182, 189
cyclohexylammonium salt of PEP 269
Cycocel (see CCC)
cytokinesis 95-97

tissue culture 95-105
TLC (see thin-layer chromatography)
tobacco 68, 70, 74, 96, 97, 101
tomato 153, 245
tonoplast 5, 11
trace elements 255
transketolase 67
translocation 69, 277-289
transpiration 277, 315-316
triazines 216, 218
2,4,5-trichlorophenoxyacetic acid 216
triple response to ethylene 204, 212
Tris ADP 269
__ buffer 85, 269
Tris-glycine buffer 85
Triticum aestivum 68, 219, 222
tropisms 108, 303
true dormancy 108, 131
tuber formation 108, 243
turgor pressure 13-14
Tween 20 133, 166, 169, 216-219, 228, 232

univalent cations, activation of enzymes by 267-276
ureas 216, 218, 325, 327

vacuolar membrane 5, 11
vacuole 5, 11
vanadium 257
vegetative photoperiodism 108, 243
vermiculite 123, 126, 155, 158, 164, 169, 193, 205, 208
vernalization 163
vitamins 97, 98

Warburg apparatus 71
__ effect 68-69
water potential 12-14, 315-316
__ relations 11-22, 315-324
wheat 68, 219, 222
white ash 121-129
Wisconsin No. 38 tobacco 96

Xanthium strumarium 192, 193, 244, 245, 246, 249-250
xanthophylls 35, 36

Zea mays 68, 70, 74, 216, 219, 245
zinc 255

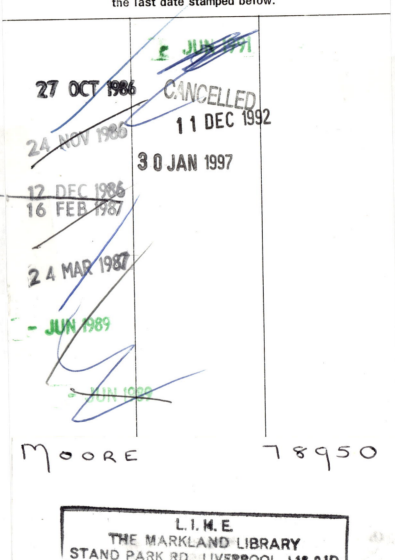